150049519X

AF616181

THE ENGINEER'S CONSCIENCE

THE ENGINEER'S CONSCIENCE

M. W. THRING

Northgate Publishing Company Ltd
London

First published 1980

ISBN 0 85298 433 2 Hard bound edition
ISBN 0 85298 463 4 Limp edition

This book has been printed entirely on recycled material and the text was composed by processes which conserve material resources and are non-polluting.

Printed in Great Britain by Adlard & Son Ltd, Bartholomew Press, Dorking, Surrey
Bound by Western Book Co., Ltd., Maesteg, Glamorgan

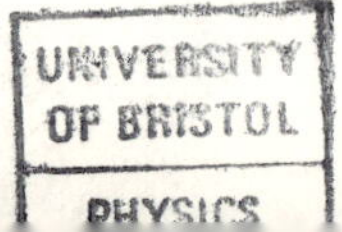

CONTENTS

(1) The only humane way to halt the rise of world population is to give all mankind a decent standard of living and an adequate education. This must be done within one generation so that the world population levels off at about twice the present figure, i.e., at 8000 M.

(2) To avoid the Third World War world tensions due to the jealousy of poor countries of the wasteful life style of the rich must be relaxed by bringing the per capita use of the earth's resources (e.g., fossil fuels) to about the same level all over the world.

(3) To avoid excessively rapid depletion the world average per capita consumption must remain at about the present level so that with twice the world's population we only use twice the present resources. This means the rich countries must reduce their per capita consumption to about one-third of the present level, and the poor countries rise to this figure.

(4) Only those engineering developments are permissible in the long term which can be made available to all 8000 M people. Everything else is a 'rich man's toy'.

Thus, it is clear that only a radical change of policy in the rich countries can lead to a stable civilization. *It will not happen automatically*, and it requires a complete change of the criteria of success in life, so that people struggle in life for creative self-fulfilment, rather than for possessions beyond all that are necessary for such a life. It may be possible to achieve this change when enough people see clearly that the end result of status symbols is the total destruction of our civilization, as all previous ones have been destroyed.

The rich 30 per cent of the world's population uses 80 per cent of the fossil fuel energy. The average per capita energy consumption in the rich countries is 5–6 TCE/person–year (tons of coal equivalent per person per year) and in the poor countries, 0·5 TCE/person–year, making the world average just under 2 TCE/person–year. A man's food is equivalent to 1/6 TCE/person–year.

It is essential to look at the energy supply on a worldwide basis since this gross disparity creates world tensions that must lead to world war, as the very limited oil resources of the world are used up without the poor countries having their share. Oil is uniquely suitable as a cheap natural material for transport, agriculture, and petrochemicals. Substitutes made from coal shale or biomass will cost several times as much.

Coal is available in very much larger quantities but its world use cannot rise to much more than twice the present without carbon dioxide causing problems.

In addition to the well-known dangers from radioactivity, nuclear energy cannot be regarded as a major contributor to the overall problem of supplying enough energy to 8000 M people, since the resources needed to build large (e.g., 1000 MW) central power stations (with a lifetime of only thirty years) with an adequate distribution system could not possibly be found.

The necessary steps to a stable world energy use are as follows:

(1) During the next twenty to thirty years the rich countries have to come down, and the poor countries to come up, to an energy consumption about equal to the present world average figure of 1·8 TCE/person–year. This means, for example, that Britain's total energy consumption must fall by 3–4 per cent every year, which can be done, without any sacrifice of real benefits to the individual, by applying known methods of fuel economy and waste elimination.

(2) The premium fuels (oil and natural gas) should be used solely for the purposes for which they are uniquely suited, and must be used with great economy even for these purposes, so that they are also available for the poor countries. Coal and refuse should be used for all stationary heating purposes, electricity generation, and trains.

(3) Electricity should never be used for low-temperature heating (except with heat pumps); pass out heating from thermal power stations should be used for these purposes wherever possible. This could halve the primary fuel used for electricity generation in Britain, with the same end result.

(4) Methods of bringing coal (and other minerals) to the surface without men going underground (telechiric mining) should be developed so that all the coal under land and sea, and at any depth, can be won.

(5) While the fossil fuels last we should invest enough in renewable systems (e.g., solar heaters and power generators, and wind generators) so that our descendants can have what we have now, even when we have exhausted the fossil fuels. These will probably be in small and medium-sized units (e.g., 3–5 kW), supplying a village, factory, or small town.

Oil will be relatively much more expensive in the next century but it is likely to remain the most convenient fuel for air, sea, and road transport because it is carried in light tanks and reacts with fourteen times its own weight of air which is not carried.

The private car consumes 70 per cent of the oil used for transport, and transport consumes about 30 per cent of the total oil used. Only a few very rich people in the underdeveloped countries have cars, whereas in the rich countries very many families have one or more.

Other problems which must be solved are traffic jams, pollution (lead, carbon monoxide and hydrocarbons especially) and noise. The solutions require a reduction of present top speeds, much more convenient public transport, better design against pollution and noise, and more use of the bicycle. If private cars are feasible at all they will have to do 100 miles/gal, last fifty years, carry two or more people on all regular journeys, and use unleaded fuel.

Rail transport which can readily operate on coal or solid biomass is likely to play a much bigger role in the future all over the world.

Communications research will have to be orientated towards mass educational aids, reduced conference travel (TV links) and elimination of wood for newsprint.

It is essential to the establishment of a stable Creative Society in the next century that we find how to grow enough food to give 8000 M people:

(1) Energy (8–10 MJ/day), protein (1·5 g/day–kg body weight), vitamins, roughage, and trace elements necessary for full bodily health.

(2) Tasty and attractive meals.

(3) As far as possible preservation of traditional food habits.

The developed countries can grow more than enough food for their own needs but this is based on the use of exhaustible fossil fuels for fertilizers, chemicals and machines and for food processing. The seas are being depleted of fish and the deserts are growing. Cutting down forests is causing floods and soil erosion. Dams are reducing the flow of fertile silt.

In conclusion:

The rich countries *must* devote enough resources to providing the poor ones with the means to grow their own food.

In the twenty-first century the rich countries will have to take most of their protein in vegetable form. The rich countries will have to grow almost all their own food, return to mixed farming, concentrate much more labour on the land, and recycle all compost and city night soil. The leaf fractionation process will have to be extensively used in both rich and poor countries.

cardiac pacemakers and artificial kidneys, and with new surgical tools (lasers and sewing machines).

Sceptrology (the technology of 'crutches') can provide stair-climbing wheelchairs, battery operated carriages for limbless children, standing and stair-climbing aids, and devices for blind and handicapped people living alone.

Devices for the totally deaf may even enable them to recognize speech by electric pulses to the skin.

The engineer has contributed very valuable devices for whole body scanning and intensive care.

Architects need to give much more thought to the use of buildings by cripples.

CHAPTER 9 The engineer's responsibility 222

The engineer is necessarily faced with moral questions in his work. For example:

(1) Short-term or long-term benefits? Fast cars now or oil conservation.

(2) Benefits for the few, coupled with smaller disadvantages for many (e.g., noise and pollution).

(3) Should he devote his skills to weapons of war?

He must develop his own conscience so that he can take a long-term worldwide view of the consequences of his work. Only the engineer can prevent people saying 'It's all right to do what we want now, the engineer will solve the problems in the future'. This applies particularly to energy and the limited fossil fuel resources.

Many individual writers, and many conferences, have considered these problems and come up with essentially the same conclusions about the need to be aware of the harmful social consequences of much of modern technology.

If it is accepted that the long-term objective of achieving a stable world society for our grandchildren in the twenty-first century takes priority over all short-term objectives, then clearly for the engineer to design weapons of destruction in peacetime is totally wrong, for it can only lead to their use, as well as to the moral degeneration of the designer. To spend the same skill and resources on helping the poor countries would make a very much greater contribution to the peace of the world.

To young people, thinking of becoming engineers, architects or applied scientists, my advice is:

'Qualify as well as you can; get inside the system (but not war industries) and keep your conscience active as you follow your career. As you rise in the system you can exercise more influence to take a long-term worldwide view of problems.'

To my wife, who has helped me in
every possible way for nearly forty years.

Conscience: an individual's system of accepted moral principles; the subjective sense of right and wrong, especially as applied to a particular act. (*Penguin English Dictionary*)

CHAPTER 1

The effects of the Industrial Revolution on human life

1.1 HIGH TECHNOLOGY, RADICAL TECHNOLOGY AND HUMANE TECHNOLOGY

Anyone now at school or university will expect to have an active career for the next forty years. Suppose we could construct a time machine, as postulated by H. G. Wells, and set the journey's end to the year 2020; what would we find when we come out? Fortunately the future is not fully predetermined but can be influenced by man's common sense, otherwise we would be certain to find our whole civilization in ruins. There can be no doubt whatever that all the signs are present of our civilization destroying itself in one of the classical ways by which all previous civilizations have destroyed themselves. Among the causes leading to disaster one can mention:

(1) an increasing proportion of engineering effort devoted to weapons,

(2) accidents and pollution due to cheap engineering,

(3) short-sighted wastage of scarce resources such as oil,

(4) the concentration of engineering effort on prestige projects such as the space programme, Concorde, and nuclear power, when more than half mankind have not had any of the real benefits of engineering (e.g., sanitation, hygiene, land drainage, irrigation, comfortable houses).

The engineer alone cannot avert the potential disasters of the Third World War, world wide famine, and pestilence, or a complete breakdown of law and order until every individual is prepared to kill his neighbour in self defence, but the examples given already show that the use of engineering skills for inhumane and short-sighted purposes is the principal factor in human behaviour leading to disaster. Thus the development of *the conscience* in the engineer is a most essential requirement if our civilization is to emerge from the shadow of disaster and reach a sunny upland in which all humanity can live in peace and self-fulfilment.

The engineer with a conscience considers the long term effects of what he does on the life of all human beings and on the atmosphere.

High Technology is the extrapolation of the technological trends of the Industrial Revolution far beyond the point at which they give to the inhabitant of the developed countries all he needs for a fully satisfactory life. This overshoot has occurred particularly in the thirty years since the Second World War. It has been caused basically by the work of the engineer as was the Industrial Revolution in the first place (see section 1.2).

Examples of High Technology in civilian life.

(1) The increase of speed far beyond that necessary for real human needs: supersonic passenger aircraft where more time is spent on the few miles of ground journey than on the thousands in the air; 150 mile/hr trains; hovercraft and cars capable of 250 mile/hr.

(2) The increase of family expenditure on short-lived capital goods which replace traditional skills and self-reliance.

(3) The increase in size and complexity of factories and the use of more and more automation and more robots, computers, and microprocessors which cause unemployment and make the remaining jobs more and more specialized and so less satisfying to human beings.

(4) The increasing dependence on large central electricity power stations with all their problems of distribution and breakdown.

(5) Cities grow larger and larger and commuting journeys correspondingly longer. The tide is already turning against the skyscraper office and high-rise flat as ordinary people revolt against their inhumanity. These are a consequence of the engineer's inventions of the lift and the steel or concrete building frame.

(6) Vast sums are spent in the developed countries on space rockets, nuclear power stations, and heart transplants, while money is not available for schools and hospitals or for helping the underdeveloped countries.

In the military field examples of High Technology are the intercontinental ballistic missile, the homing missile, and the megaton nuclear explosive.

Some particular examples of the extremes to which High Technology takes us are:

(1) The proposal to use space rockets to fabricate pure materials under zero *g* conditions ('Skylark, Smithy in Space', *New Scientist*, December 1977, p. 556).

(2) 'A baby born in Space'. A Russian plan to achieve a human birth

in space was described by *The Times*' Moscow correspondent, 18 September 1978.

(3) The Paris–Strasburg motorway is so soporific that brightly painted totem poles have been erected along it to keep drivers awake.

It follows that the engineer must accept a large measure of responsibility for both the good and bad consequences of the Industrial Revolution and of High Technology. The engineer is no more entitled to say 'I was ordered to do that by the politicians and so I am not responsible for the consequences', than was the commandant of Belsen, for the engineer is the only person who can foresee the physical consequences of the developments that he plans to make. The primary aim of this book is to try to help the engineer to develop his own conscience to the point where he makes use of all the freedom he has to choose, from those available, which organization he shall be employed in, and what he is prepared to do for that organization.

The book is naturally a statement of my own views and conclusions (backed up as far as possible by facts and quotations from other authors) about the way in which an engineer with a conscience can use his skills to improve the life of the ordinary individual. Thus, I shall have to trespass on other areas by asking such questions as 'What does constitute a fully satisfactory life for an individual?' and 'What is the purpose of life?', questions which are normally regarded as being outside the concern of engineers. In these matters of conscience there are no black and white, clear-cut conclusions or definitions; they are highly subjective, as, for example, is the definition I shall give in Chapter 3 of what I mean by Quality of Life. Nevertheless, the engineer must consider such subjective matters when, for example, he is concerned with the question of whether it is more important to make motor cars with higher acceleration or with greater safety.

I do not suggest you believe what I say, but I want to arouse thought on these important matters. Buddha said 'Believe nothing anyone tells you, not even what I tell you' and this must apply far more strongly to the writings of an ordinary person. The 'do-gooder' who imposes his views by force on other people, in the long run does much more harm than the person with absolute power who is a straight-forward crook*.

* This view was expressed by Professor Pusic at the Siena Seminar of the Acton Society, 1977. If people with absolute power had fully developed their own love of humanity and their power to think without prejudices or selfishness before they came to power, this would not apply, but unfortunately, as Carlyle pointed out in his study of the French Revolution, the ability to reach a position of power is not at all the same as the ability to exercise power wisely and well.

Thus, I am not suggesting that everyone's conscience will arrive at the conclusions that I express here, but I am suggesting that *every engineer has the responsibility to think these matters out for him or herself.*

I will mention three examples of areas where the impact of High Technology on human life is extremely worrying to large groups of thoughtful people.

The first of these is the extraordinary success and skill of the engineer in developing weapons of destruction of people and their homes and work places. This is clearly a very complex matter. For example, most people would agree that Hitler was like a mad dog which had to be stopped by force. On the other hand, we are not in a situation of that kind at the moment and yet, according to the Stockholm Peace Research Institute, (SIPRI) some 40 per cent of all the scientists and engineers of the world are now working directly or indirectly on weapons of destruction. It is also certainly true that the vast expenditure on nuclear power generation would never have been incurred if there had not been the military 'need' for nuclear explosives, and the development of space research in Russia and America would not have taken place without the 'need' for intercontinental ballistic missiles. There is absolutely no doubt that a brilliant engineer who devotes many years of his life to the development of successively more powerful weapons of destruction becomes dehumanized and his conscience totally destroyed, so that he calculates the destruction of human beings and their homes with no feelings of pity at all. Moreover, he may well become a 'hawk' simply because this gives him more money for the development of the form of high technology by which he is fascinated. Thus, the escalation of weapons absorbs a higher and higher proportion of the country's income at the expense of medical services and the educational system. It has even reached a point where it is almost impossible to destroy the weapons which have been built up in a safe way.

The second major area in which the engineer's conscience should become more and more involved, is the needs of the less well developed countries, and particularly those areas where people are at near starvation levels or where disasters occur. Problems like the fact that the Green Revolution was based on oil, which can no longer be afforded in the poorer countries, while it is being used in ways which can only be described as extravagant and luxurious in the developed countries, should weigh on his conscience.

The third problem for the engineer's conscience is the damage which our machines are doing to the environment, and to people's freedom of lifestyle, by pollution of air, water, and land—noise, destruction of wildlife, covering agricultural land with concrete valleys for reservoirs—and in the long run replacing the beauty of the earth with dereliction. On a short-term economic basis all these things are desirable since it is expensive to clean up our messes and our economic system demands a perpetually rising standard of living. There is a deep feeling of revulsion in the consciences of many people, particularly the thoughtful members of the younger generation, against these evil consequences of engineering. This revulsion leads, in many cases, to rather extreme actions such as violence, the setting up of small communities isolated from society, and the development of Radical Technology*. It also leads many thoughtful young people to eschew engineering as a career.

Centres have been set up in America, in Britain and other countries where people try to live independently of the fossil fuels, nuclear energy and the big factory and much valuable work is being done. However, it is an unfortunate fact that there are now so many people in the world that a return to the technology which existed before the Industrial Revolution could no longer provide even the necessities of life, such as a full diet, comfortable homes and clothing, for all these people. Moreover, there are many things which have come from the Industrial Revolution which improve home life enormously, such as comfortable, water-tight homes with glass windows and electric lighting, the whole of hygiene and sanitary engineering, the instruments and drugs of the medical profession, and systems of travel and communications. Thus we cannot go back to the pre-Industrial Revolution level of technology. Indeed most of the Radical Technology communities end up with a compromise, as, for example, when they use a windmill to drive an electric generator which has to be made in a large factory. Some things which have to be made by mass production such as the refrigerator or the vacuum cleaner, are very positive contributions to human life and if they are made to have a long life are not excessively resource consuming. I think one is forced to the conclusion that genuine Radical Technology is not a solution to our problems unless the world population were to go back to the level it was at before the Industrial Revolution, and the real advantage to human life of the good aspects of the Industrial Revolution were to be forgone.

On the other hand, the further development of High Technology is clearly leading further and further away from a satisfactory human existence. High Technology is based on the assumptions that some people can have a perpetually rising standard of living, and travel at perpetually increasing speeds. This requires an exponentially rising per capita energy

* Godfrey Boyle and Peter Harper, *Radical Technology*, 1976 (Wildwood House, London).

consumption (+2–4 per cent p.a.) with a corresponding rise in the use of metals, fertilizers, plastics, and other sophisticated materials. It also implies more and more dependence on electricity to run lifts, freeze food, and heat water, and on cars to get us long distances to work. Above all it is based on the idea of perpetual economic growth of an exponential character. Thus we are frequently told, as engineers, by the politicians and the economists, that we must produce a 4 per cent per annum increase in productivity (output per hour of a man's work in a production factory) to maintain a perpetually rising standard of living. This in turn is based on an exponential increase in the use of fossil fuels and raw materials of other kinds. It is an inevitable consequence of this perpetual striving for growth and cheapness that standards of quality and safety are not given sufficient priority and therefore they actually fall as the engineer is forced to cut corners.

Humane Technology may be defined as 'Engineering as though people mattered', using a phrase borrowed from the sub-title of Schumacher's book* *Small is Beautiful*, which is *A study of economics as if people mattered.* 'People' in this connection must mean ordinary people and not just those who can afford to pay a high price for the products of engineering. Humane Technology or Humane Engineering is thus the application of the intelligence and skill of the engineer to the problem of providing all the people of the world with a sufficient standard of living for them to be free to develop fully all their latent human possibilities. This must be on a basis of an equilibrium economy in which man lives in a steady-state relation with his environment. Moreover, as we shall consider in Chapter 7, all normal people should be able to earn this standard of living by doing an interesting job which occupies their talents well and which takes up a substantial fraction, but not all, of their best life's energy.

The problem which we have to try to answer in this book is how it is possible to develop Humane Engineering to satisfy this condition for 8000 million people within the limited resources of the earth (especially land, fresh water, clean air, and minerals), and at the same time to retain sufficient wilderness and wildlife so that man does not totally destroy the natural ecological system of the planet. Pollution of all kinds must be reduced to a level at which it neither has any detectable long-term consequences on human, animal or plant life, nor is directly noticeable.

Many books have been written on this subject and many societies formed to study it (see Appendix at the end of this chapter). The particular

* E. F. Schumacher, *Small is Beautiful, A study of economics as if people mattered*, 1974 (Abacus, Tunbridge Wells, Kent).

function of this book is to try to study the subject from the point of view of the engineer with a conscience, and to give that information which is necessary for young people entering the engineering profession. They need this information to understand the vital decisions affecting our civilization which will be taken during their career, and to enable them to play as positive a role as possible in ensuring that the choices made are the right ones.

Many of the people concerned about the future of a technology-based civilization are like myself, University Professors, because our job of helping young men and women to prepare themselves for a forty-year professional career gives us an enormous responsibility to look ahead.

In her book *Philosophers of the Earth** Anne Chisholm describes discussions with some eighteen ecologists, all of whom are seriously concerned with the effect of our present mode of life on the ecology. Another book closely concerned with this matter is *Can Britain Survive?*† The following is quoted from the preface by E. J. Mishan.

> A larger, more informed, concern is necessary to bring pressure on governments to move away from their fixation with the policy of headlong economic growth towards a policy based on ecological stability and on a number of criteria of social welfare; to move, that is, from commitment to economic growth to a commitment to quality of life.

1.2 ENGINEERING FEATURES OF THE INDUSTRIAL REVOLUTION

The causes of the Industrial Revolution were manifold and the reason that it started primarily in Britain were political and social as well as technological, but the fundamental factor was technological advances in metals and machinery. *The Protestant Ethic and the Spirit of Capitalism*‡ suggested that in Britain the Industrial Revolution was largely the work of a single group of people, the non-conformists for whom working hard and constructively was an essential part of their religion. A. J. Meadows (*The Industrial Revolution* §) suggested that it occurred in England and not in France, which had three times the population, because the medieval system of organizing labour, which restricted movement about the country, had virtually disappeared in Britain and the British deposits of iron and coal lay close together and were fairly easily worked. The 1958 *Encyclopaedia Britannica* article also points out that Britain had experience of

* Anne Chisholm, *Philosophers of the Earth*, 1972 (Scientific Book Club).

† E. Goldsmith, *Can Britain Survive?*, 1972 (Tom Stacey).

‡ Max Weber, *The Protestant Ethic and the Spirit of Capitalism*, 1930 (Allen and Unwin, London).

§ A. J. Meadows, *The Industrial Revolution*, 'Science in Society' leaflet, 1978.

foreign trade and relations with the New World, but again stresses the fact of plentiful coal in Britain. Desmond Lee* says that the Industrial Revolution did not come in the ancient world largely because manual labour was done by slaves and was regarded as degrading so that little interest was taken in improving these tasks. He stresses that the two essential bases of the Industrial Revolution were the new source of power, steam, and the new fuel, coke, which made it possible to make plentiful iron.

The Industrial Revolution is generally taken as beginning around 1750 and coke was first used to replace charcoal for making iron by Abraham Darby around 1720, when the total annual production of pig iron in Britain was only 25 000 tons. At that stage the charcoal burners had largely depleted the primaeval forests of Britain. By 1788 we were making 68 000 tons of pig iron and by 1839 1 347 000 tons. In the middle of the nineteenth century Bessemer and Siemens developed their steel-making processes which enabled us to replace puddled wrought iron and crucible high carbon tool steel with tonnage steel. Thus, the material was available for the really big expansion of the Industrial Revolution.

The other key process was the development of steam power. James Watt patented the separate condenser for the beam engine in 1769 and developed systems whereby the beam engine could be used to produce rotary motion instead of just working reciprocating pumps. The steam engine was applied to work on textile looms in 1785. George Stephenson built the first really successful locomotive engine which required high pressure steam, because it discharged at one atmosphere, for the Stockton and Darlington Railway in 1825. This led to the fast expansion of the railway system which largely replaced the earlier canal system for goods transport; by 1848 there were 5000 miles of rail track in Britain. Since then there have been enormous developments in the production of power, the stationary internal combustion gas engine being available at the end of the nineteenth century, driving groups of machines with belt transmission. The transmission of power by means of electricity started specifically with the work of Edison about the turn of the century and has developed up to the modern grid system with steam turbines of several 100 MW.

1.3 THE GOOD CONSEQUENCES OF THE INDUSTRIAL REVOLUTION

Everybody knows that the Industrial Revolution has improved human life immensely, but it is very important to distinguish between the real

* Desmond Lee, *Science, Technology and Society in the Ancient World*, 'Science in Society' leaflet, 1978.

improvements, which raise the quality of life of the individual, and those which are merely status symbols, which he has to be persuaded to buy by expensive advertising, and the concept of 'keeping up with the Jones's'. For the present purpose we shall take the quality of life as meaning those things which enable people to develop their bodies, their minds, and their feelings, without restrictions due to shortages or compulsion. These may be illustrated as follows.

1.3.1 *Food*

By producing agricultural machinery, fertilizers, pesticides, high yielding crops, and good transport systems, the Industrial Revolution has given plenty of nourishing and varied food of good appearance to most people in those countries which have had the benefit of it. There are many question marks; for example, the use of preservatives with doubtful effects, the feeding of babies on powdered milk, and the effects of pollutants such as lead getting into our food, which will be considered in Chapter 6. Nevertheless, undernourishment has been eliminated in these countries.

1.3.2 *Homes*

It has proved possible to build homes at such a price that a person with a reasonable job in a developed country can afford a comfortable, convenient, draught free, well decorated home with good windows and heating. They can also afford to furnish it with comfortable furniture, even though it is often not very long lasting. The standard of comfort is higher than that of princes or millionaires 200 years ago.

1.3.3 *Health*

The expectation of life has increased vastly, particularly because infant mortality has been so reduced. This is partly the result of the development of specific drugs but much more the result of preventive medicine in the form of hygiene and sanitation, which are consequences of engineering. New problems have arisen in the health field, dealt with very thoroughly by Ivan Illich in *Medical Nemesis*, but certainly hygiene and sanitation and other preventive medicine steps are of immense value to human life.

1.3.4 *Education*

The engineer has made a very real contribution to the improvement of education, first by improvements in printing and cheaper books and, more recently, by communication systems such as films, television and tape recorders, which can give access to information about the whole world, and its problems, to all people. Machines cannot replace good teaching by human teachers who have a direct relationship with the students, which makes possible stimulation and individual feedback, but they can be of immense help in widening the scope of the material available and making it much more vivid. As with wealth, there is a serious fault in our present educational system due to lack of human commitment. The students are not taught to think for themselves and there is a lack of education of the emotional brain and the skill of the hands, but the engineer's contribution has been of undoubted real value. There are also problems due to the dominance of the financial motive (e.g., audience ratings and newspaper sales competitions), but the inventions themselves are potentially of enormous value.

1.3.5 *Leisure*

The enormous increase in the productivity of a factory or farm worker from an hour's work, which is somewhere between 10- and 100-fold, compared with what it was before the Industrial Revolution has enabled people to obtain their living and still have enough energy and leisure time to have creative hobbies, healthy exercise, and take part in sporting recreations to a fully adequate extent. The problems of unemployment due to this proportion of leisure becoming too great are discussed in detail in Chapter 7.

1.4 BAD CONSEQUENCES OF THE INDUSTRIAL REVOLUTION

If we take as examples of products of the Industrial Revolution the motor car and the lorry, we can make a balance sheet of good and bad consequences. The good consequences include family door-to-door mobility, specially convenient with children and awkward luggage. This is accompanied by warmth and dryness in all weathers. The lorries have given us a world-wide integrated food, fuel, and manufacturing system. The disadvantages of cars and lorries include nerve damage, noise, traffic jams, accidents (specially in icy or foggy weather), pollution of the air (especially lead and carbon monoxide and partially burned hydro-carbons which may be carcinogenic), lack of exercise, and they also contribute to lengthy and inconvenient commuting journeys and human rootlessness.

Similar balance sheets can be drawn up for many of the other engineering 'blessings', for example the lift is a real boon to cripples but has led to the disaster of the high rise flat and the uselessness of skyscrapers during power breakdowns. The computer has been hailed as the greatest blessing for mankind but as far as ordinary people are concerned it has reduced the proper human service given to them by banks or Civil Service organizations. Even the electric grid system has led to lines of pylons marching across the countryside and great plumes of water vapour from the cooling towers and sulphur dioxide from the chimneys. In this section I shall list briefly some examples of the bad consequences of the Industrial Revolution without attempting to cover the subject comprehensively.

1.4.1 *Damage to the environment*

Many books have been written on the subject of Air Pollution*. Under this heading one problem that needs special mention is the lead put into the atmosphere by the exhausts of our cars, which will be considered in more detail in Chapter 5. The Danish Ministry of Environment suggests that people living near busy roads may be exposed to so much lead from the atmosphere that 'proposed and approved hygiene limit values for airborne and dietary lead are equalled or exceeded'†. A campaign against lead in petrol has been set up‡.

Another problem in air pollution is the use of chlorofluorocarbon (CFC) aerosol propellants. The Swedish Ministry of Agriculture is banning these in Sweden because it is considered possible that this reduces the ozone in the upper atmosphere and it is this ozone layer which prevents life-damaging ultra-violet light of a particular wavelength from reaching the earth§.

A third example is sulphur dioxide from the combustion of fuels, which has been shown to cause damage to vegetation and fishes as a result of sulphuric acid deposition on those countries which have an acid soil||.

It is almost impossible to pinpoint any particular air pollutant as the main factor in causing human deaths and ill-health, although arsenic, asbestos, and coal tar have been proved to be carcinogens, while sulphur dioxide and carbon monoxide are known health hazards. However, most of the medical statistics can only compare generally polluted air with

* For example: R. S. Scorer, *Pollution in the Air*, 1973 (Routledge and Kegan Paul, London). Werner Strauss (Ed.), *Air Pollution Control*, 1971 (Wiley, New York). W. L. Faith and Arthur A. Atkisson, Jr, *Air Pollution* 2nd edition, 1972 (Wiley, New York). Paul R. Erhlich and Anne H. Ehrlich, *Population Resources Environment, Issues in Human Ecology*, 1970 (Freeman, San Francisco).

† 'Lead astray by roadside vegetables,' *New Scientist*, 15 December 1977, p. 686.

‡ 'The case against lead in petrol', *New Scientist*, 11 August 1977, p. 348.

§ 'Mixed responses to aerosol propellants'. *New Scientist*, 1978.

|| 'Sulphur pollution across boundaries', AMBIO, February 1972, p. 15.

unpolluted. Lave and Seskin* studied the long-term effects of living in polluted atmosphere. They obtained statistics from fifty-three boroughs and related mortality rates to a solids deposition index. Thirty-nine per cent of the bronchitis mortality rate was explained as the effect of this index; a 10 per cent decrease in the deposit rate would lead to a 7 per cent decrease in the bronchitis death rate—a smoke index gives an even better correlation—cleaning the air to the best value would reduce the bronchitis mortality by 70 per cent. Similarly, a regression on lung cancer in England and Wales showed that if the quality of air in all boroughs was improved to that of the best, death rates by lung cancer would fall by between 11 and 44 per cent. Rates of mortality in part of New York State due to stomach cancer are more than twice as great in areas of high pollution as in areas of low pollution.

London postmen have 25–50 per cent higher rates of incidence of severe respiratory disease than small town postmen, while 20–35 per cent of absence due to sickness in London bus drivers could be ascribed to fog. Infant death rates in Nashville, USA, showed the highest correlation with atmospheric concentrations of sulphur trioxide. Lave and Seskin conclude that 4·5 per cent of all economic costs associated with morbidity and mortality would be saved by a 50 per cent reduction in air pollution in major urban areas.

Noise, especially from road traffic and aircraft, is one of the most subtle but really damaging consequences of the Industrial Revolution. This will be discussed in more detail in Chapter 5. There have been just as many damaging consequences to water, that is rivers, lakes and the sea. DDT and other pesticides and excess fertilizers are washed from farm land into the water and produce damage to marine life which may eventually finish up as human food. In particular, nitrates, resulting from nitrogenous fertilizers, are appearing in rivers and in ground water sources, and these can cause the blue baby syndrome in infants and gastric cancer in adults. In 1976 the Anglian Water Board had to provide bottled water for babies because half the water exceeded the 50 mg/l level†.

A study by the European Commission showed that a large number of toxic or carcinogenic substances are discharged into European water from farms and factories and that the addition of chlorine can make some of these substances even more harmful‡. Mellanby§ says 'I don't

* Lester B. Lave and Eugene P. Seskin, 'Air pollution and human health', *Science*, 1970, **169,** 273.

† M. Ottaway, 'Farming threat to clean water', *Sunday Times*, 19 February 1978, p. 18.

‡ European Commission, 'Euroform' No. 5/78.

§ Anne Chisholm, *Philosophers of the Earth*, 1974 (Scientific Book Club), p. 77.

think there is any doubt that the most serious pollution in this country is in water, from industry and from insufficiently treated sewage'.

In 1969 millions of fish in the Rhine were killed in three days by a large discharge into the water from a chemical factory*. Pollution of the Mediterranean is exceptionally severe as 90 per cent of the sewage from 120 coastal cities flows untreated into the sea and cannot be disposed of rapidly by it†. Viral hepatitis, dysentery, typhoid, and poliomyelitis result in the neighbouring areas. The concentration of mercury in tuna fish averages three times higher than in the Atlantic and is above the safe limit.

More than one-eighth of all the world's oil spills occur in the Mediterranean. Spilled oil is 108 mg/m^2–yr compared with 17·45 mg/m^2–yr in the North Atlantic. High levels of zinc are found in the Mediterranean area, and along the flower-growing coast increased levels of DDT are found in mussels†.

The per capita use of water in the United States of America is doubling every forty years, the sewage produced being about 150 gal/person–day, while the industrial waste water is 2·6 times as great as the domestic. Phenols, cyanides, sulphates, manganese, nitrates and nitrites, chloride, fluoride, copper, lead, zinc, chromium, sulphide, mercury, complex organic chemicals, and arsenic are among the pollutants which are being discharged by factories into rivers or sea water. A sixteen-acre housing estate at Niagara Falls has been evacuated and 234 families rehoused because drums of chemical dumped in a disused canal and covered with clay had leaked eighty-two chemicals, and the area had the country's highest average for miscarriages and deaths from leukaemia and cancer. Arsenic is a cumulative poison which builds up slowly in the body and the US Public Health Department's recommended maximum in drinking water is 10 ppb. Angino *et al.*‡ have found up to 8 ppb in the Kansas river and up to 41 ppm in certain detergents and pre-soaks. Arsenicals are also used as farm insecticides.

Tolley and Reed§ have sampled the tap water of forty-three English boroughs and found that 13 per cent were contaminated with lead to more than 0·11 mg/l, at which figure the World Health Organization recommends that water should be rejected for public supply.

* *Ibid.*, p. 96, Donald Kuenen.

† Jeremy Bugler, 'Mediterranean countries come clean', *New Scientist*, 5 January 1978, p. 4.

‡ E. E. Angino *et al.*, 'Possible danger and pollution hazards', *Science*, 1970, **168,** 389.

§ D. Bryce Smith, 'Land pollution; a growing hazard to health', *Chemistry in Britain*, February 1971, **7,** No. 2, 54.

Twenty years ago fishermen, their families, cats, and the local sea birds at Miniamata Bay in Japan, suffered severe structural damage to the brain which was traced to alkyl mercury compounds discharged by a factory into the sea and concentrated by a factor of more than 50 000 in the fish. Farmers who have themselves eaten grain seed treated with organic mercurial fungicides or have fed it to their pigs have had severe poisoning*. Alkyl mercury can cause congenital mental retardation†. It is quite certain that great care must be taken whenever there is any possibility of mercury compounds reaching fishes or animals which will eventually be eaten by man.

Almost all the beaches of the world are contaminated by oil floating in from tankers. Every year or so there is a major disaster in which the tanker breaks up and thousands of tons of oil are discharged, and, although there is an international law against it, it is likely that some ships still clean their tanks on their return journeys and discharge the cleanings into the sea.

Another serious problem in regard to water pollution is what is known as thermal pollution. Conventional electricity generation from fossil fuels or nuclear energy lose about 60–70 per cent of the heat of the fuel in the form of chimney waste heat and, to a much greater extent, warm water. This warm water may be cooled by means of a cooling tower, in which case water vapour and warm air are discharged from the stack, or it may be discharged into a river or lake, which is common in the USA. Raising the water temperature from 55 °F ($\approx$12·8 °C) to 68 °F (20 °C) causes a loss of some 18 per cent in the oxygen carrying capacity of the water and this is equivalent to the addition of a substantial amount of rotting material, since the oxygen in solution is required for the oxidation of organic materials, which in turn can be lethal to fish life. A simple solution to this problem which will also save fuel is discussed in Chapter 4.

When dams are erected to produce immediate short-term benefits in water control for agricultural purposes and electric power production they do, in some cases, produce earthquakes because the land is not used to carrying the weight of water‡. The Aswan High Dam has brought another 3 M acres of land under cultivation but has upset the natural balance of the whole area in a number of ways. The sediment which used to be carried down at the period of flood is now retained in the dam and there is a serious danger that three fresh water lakes on the delta will be

* A. Curley *et al.*, 'DDT metabolism: oxidation of the metabolite 2,2-bis (p-chlorophenyl) ethanol by alcohol dehydrogenase', *Science*, 1971, **170,** 65.

† L. L. Goldwater, *Scientific American*, May 1971, **224,** No. 5, 15.

‡ J. P. Rothe, 'Fill a lake, start an earthquake', *New Scientist*, 1968, p. 75.

destroyed. The dam has also affected the fishing along the Nile and in the Mediterranean because it stopped the flow of nutrients in the river. The expansion of irrigation has also spread.*

A. K. Agarwal† says that the raising of the water in the district surrounding the Nagarfunasagar Dam in Southern India has made the soil more alkaline and hence upset the trace element balance. This has increased the molybdenum intake of people living on sorghum and caused an outbreak of a crippling bone disease among young people.

Our attitude to land is just as careless. Britain is littered with spoil heaps from coal mining and china clay pits and with derelict quarries. We even have open mine shafts in the area of the old Derbyshire lead and fluorspar mines. There are many areas of Britain, particularly in the Black Country and in South Wales, which are littered with ruined factories because it is not economically worthwhile reclaiming the land.

The tipping of dangerous industrial wastes on to the land has led to many injuries to health and life. One well-known pollutant is cyanide from plating works and if this is mixed with acid, as has happened on occasion, the deadly cyanide gas is emitted. Concentrated acids themselves have been dumped on sites, and even plutonium (which is lethal in concentrations of 10^{-10} of those of cyanide) has been found on a land-fill site in Germany. In Holland land pollution is taken so seriously that the MV *Vulcanus* steams in circles in the North Sea incinerating wastes which are considered too toxic for land fill disposal‡. There is no doubt whatever that in a truly humane society such wastes would be entirely recycled by suitable chemical treatment, avoiding all mixing.

1.4.2 *The damage caused by the Industrial Revolution to present human life*

It is no exaggeration to say that in the last 200 years the inventions of the engineer have increased the productivity of the soldier (as measured by the number of people he can kill or the number of homes he can destroy with an hour's work) by a factor of the order of 10 000. In the same period the productivity of the civilian worker has been increased by a factor of somewhere between 10 and 100. It has been demonstrated that as a result of the nuclear explosions in Japan and the atmospheric tests since then every child born afterwards has some radioactive strontium replacing

* M. Kassa, 'The River Nile ecological system', paper submitted to the UNESCO International Programme on Man and the Biosphere, 1969. Published in *Impact of Science on Society*, 1970, **XX,** No. 3, 241.

† A. K. Agarwal, 'Crippling cost of India's big dam', *New Scientist*, 30 January 1975, p. 260.

‡ Bob Keen, 'Toxic substances—wasted opportunities', *New Scientist*, 12 December 1977, p. 688.

calcium in their bones*. It is also true that if a small fraction of the nuclear explosives which are in the warheads of intercontinental ballistic missiles at the present moment were to be discharged in war, the resulting radioactivity would drift round the world and all babies of humans and animals would stand a very high chance of being born abnormal thereafter. Yet engineers go on designing, and governments building, more of these horrors as a result of the argument that they cannot let the other side establish a lead. It is true that we have gone for more than thirty years since the last world war with only minor wars in which the most terrible weapons have not been used. On the other hand, the developed countries are busily exporting nuclear power stations and every kind of sophisticated military machine to any country that can pay for them. Any thoughtful person must realize that this situation can easily lead to a holocaust.

The other major problem is closely connected with the risk of war. It is the complete failure of the developed countries to make a substantial contribution to solving the problems of the underdeveloped countries to give them the benefits of the Industrial Revolution. This, of course, creates a dangerous situation, not because the underdeveloped countries can afford the weapons and the fuel to attack the developed countries, but because the world tensions created by the jealousy of the 'have nots' for the 'haves' (who are wastefully using up the world's limited resources) is constantly causing small wars, any of which could rapidly escalate to include the great powers.

There are two reasons why the rich countries have failed to help the poor countries.

The first reason is that they are not prepared to make any real sacrifices of their own standard of living in spite of the fact that their own future must depend on it. This is illustrated by a statement that the US aid given to India in twenty years is equal to what was spent in one month on the Vietnam war†. There is absolutely no doubt that if even a comparatively small country like Britain were to redeploy the research skill, manufacturing resources, and financial research that are spent on so-called defence in the problem of helping a country like India, they could make a major contribution in a few decades.

The second reason is that such help as has been given is of the wrong kind. As Schumacher has pointed out in *Small is Beautiful* what they need

* Lord Richie Calder, 'Hell upon Earth', p. 17, *The Environmental Handbook* (Ed. J. Barr), 1971 (A. Ballantine/Friends of the Earth, in association with Pan Books, London).

† R. S. Scorer, *Pollution in the Air*, 1973 (Routledge and Kegan Paul, London), p. 135.

is not food but the equipment to grow their own food, or fish for it, and these tools must not be the sophisticated ones requiring high energy which are suitable for a country like Britain. This problem will be discussed in full in Chapter 6 and it is one to which the engineer can make a major contribution if he is able first to study what the country needs on the spot, and then to work closely with the people in the country in developing and supplying the raw materials, such as fertilizers, which can put them in a position to be self-supporting. Much good work is being done by engineers such as those in the Intermediate Technology Development Group, but the amount of financial support received is far too small to make any real contribution to the problem.

Another aspect to which far from sufficient thought has been given in engineering design is the question of accidents. It is true that some road accidents can be blamed on people's vanity, in determining to overtake, or their carelessness, in not allowing for the danger of slippery or foggy roads, etc., but nevertheless, as Ralph Nader described in *Unsafe at any Speed*, the motor car is designed in such a way that it is exceedingly dangerous. He has even described the modern American motor car as 'an enormous suit of armour designed to protect the driver when the inevitable crash occurs'. Cars are given the names of weapons of war and are advertised on the rapidity with which they can reach a speed far above the legal limit. Safety is only improved after accidents have shown the weakness of a design instead of being built-in from the beginning. The same criticism may be made of domestic electrical gadgets, factory machinery, fishing vessels, and mining machines. The machine is always assumed innocent until it is proved guilty.

In a similar way the engineer has made a major contribution to the inhuman life of people in big cities at the present time. The lift has already been mentioned, as has the lack of exercise of many town dwellers and people who commute from door to door every day. Many cities are nothing but hideous concrete ant heaps and people living in them suffer from simultaneous over-crowding, lack of privacy, and loneliness. Their food is so processed that the goodness is largely taken out of it and sold back to them in expensive packages, as in the case of vitamins and roughage.

1.4.3 *Lack of consideration for future generations*

The present economic and political system in both the 'Capitalist' and 'Communist' countries leads inevitably to short-term decisions, with a totally inadequate consideration of the long-term consequences. The usual argument is that since the engineer has been so clever in getting us to this point he can be equally clever in the next generation in solving problems. Any engineer knows that this is nonsense, because the solving of the present problems has been done always at the expense of ever increasing use of limited fossil fuels and the exhaustion of the richest and most easily won ores.

The energy problem will be dealt with in detail in Chapter 4 where it will be shown that our present policies are inevitably committing our grandchildren to a life in which most of the things we in the developed countries take for granted will no longer be available, in particular the private car and a plentiful supply of electricity, gas, and oil for domestic purposes.

Turning to food production, the high consumption of meat in the developed countries is based on the purchase, at relatively low prices, of protein-rich animal feed-stuffs from the poorer countries, while the people in these countries have not enough to eat. Moreover, our methods of farming have been based on the development of mono-cropping in place of crop rotation and mixed farming, the cutting-down of hedges to make large prairie-like fields and the use of chemical fertilizers and pesticides. L. B. Powell in an article on 'Deteriorating Soil'* states:

> The pressures exerted so relentlessly on farming since the 1947 Act have taken toll of the most basic resource of all—the soil. Much of Britain's farm land is now debilitated to a serious degree. It is becoming worn out. The notion that the soil is an inert mass which needs only a constant plastering with chemicals to give increasing yields, is one of the tragic follies of the twentieth century, and it has been encouraged by government policy.

I am doing the same wrong things on my own small farm for the same reason as other farmers do them, namely that, under the present economic system, I should go bankrupt if I did not.

Finally, it must be mentioned that those people who propose to solve the energy problem by building an accelerating number of nuclear fission power stations not only propose to tie-up a disproportionate fraction of our present high grade technological resources, but also to leave dotted around the countryside concrete mausolea which, after thirty years of useful life cannot be touched for hundreds of years. They are also proposing to produce many tons per year of highly radioactive waste which would reach red heat if embedded in the ground and which again must be prevented from contact with ground water for hundreds of years. A fully secure method of such isolation without human attention has not yet been developed.

* L. B. Powell, 'Deteriorating soil', in *Can Britain Survive?* (Ed. E. Goldsmith), 1971 (Tom Stacey, London), p. 70.

1.4.4 *Human paid work*

A very tragic consequence of the engineer's skill for developing methods of mass production and the breaking down of the complex overall task of a craftsman into thousands of separate repetitive components, each done by one individual, has been the virtual disappearance of the skill of the craftsman, working as an individual or in a small team, with pride in seeing the finished article well made and with responsibility for its good quality. Two hundred years ago a craftsman who made a coach, a windmill, or a piece of furniture, alone, or with his team, had all these factors fully developed and thus his work gave him a life of satisfaction. Even the peasant working in the fields, however long the hours he worked, had, at least, the satisfaction of direct contact with the natural process of growth from the soil, and with animals; and there was the fact that every movement required a slightly different response to the situation—there was considerable skill in hoeing weeds and avoiding hoeing the crop. Nowadays there are only about one-tenth as many people working on the land, although the skill of a good ploughman is as great as it ever was. The worker in the factory often does a job so beneath the skill of a human being that his or her only possible reason for doing it is the money, and their minds are totally divorced from their work. This is a psychologically damaging way to spend one's life because it increases the already existing lack of connection between the head, the hands and the feelings. This problem will be discussed in more detail in Chapter 7.

The other aspect of the Industrial Revolution is the danger of uncontrollably rising unemployment. The basic theory of the Industrial Revolution has been that every time the engineer increased the output of a worker by inventing new production machinery this could be absorbed by paying the worker more real money so that he and his family could buy more goods. This has worked very well up to a point, and has meant a tremendous improvement in the standard of living of the employed person in the developed countries. Unfortunately, perpetual expansion of the economic system and perpetual exponential increases of the real standard of living of the employed person must eventually hit a stop like any other growth curve and level off. There are three reasons for this. (i) Raw materials, and particularly fossil fuel supplies, are limited, both in total quantity and in ease of winning. It is a fact that when the real standard of living rises the energy consumption, per capita, rises at a slightly steeper annual rate, especially if the country concerned has ready access to a cheap fossil fuel. (ii) There are space limitations to the number of consumer products people can be persuaded to buy by aggressive salesmanship; for example almost every major city of the world is choking itself with private cars, competing with lorries and buses, so that any further increase of private cars on the road will produce less, rather than more, convenience for the users. Similarly the number of gadgets that can be squeezed into an ordinary kitchen provides a real limitation. (iii) The third limitation is the common sense of the consumer. While this varies from one consumer to another, there comes a point at which many people will resist the pressure from the advertisers to purchase the latest device or throw away their part worn object and replace it with another one.

As a result of these three factors the capability of increased production means over-production in terms of what can be sold, and thus either unemployment or fields full of unsold cars.

Both Keynesian and monetarist economies require perpetual economic growth. The engineer is told to increase productivity of the civilian worker by some 4 per cent/year, which would result in a fifty-fold increase in a century. This is absolutely impossible but even 1 per cent/year, when we are already close to the limit or have already hit it, will mean corresponding annual growth in unemployment.

APPENDIX

The following are among the societies concerned with the future of a technology based civilization.

(1) The Club of Rome, who have been responsible for the two world model predictions of the future described in the books discussed in section 2.3.

(2) The Intermediate Technology Development Group, founded by the late Dr E. F. Schumacher, 9 King Street, London. They publish the journal *Appropriate Technology*.

(3) The Centre for Integrative Studies, New York.

(4) The National Centre for Alternative Technology, Llyngwern Quarry Machynlleth, Powys, Wales.

(5) Friends of the Earth, 9 Poland Street, London.

(6) The Society for Social Responsibility in Science.

(7) The International Institute for Environment and Development.

(8) Pugwash, 9 Great Russell Mansions, 60 Great Russell Street, London. Publishes a quarterly newsletter.

(9) The Conservation Society, 12a Guildford Street, Chertsey, Surrey. Publishes a quarterly journal, *Conservation News*.

Some of the writers who have given serious and careful thought to the future of a technological society are:

Ivan Illich, *Medical Nemesis, Energy Equity*, 1975 (Calder and Boyars, London).
Paul R. Ehrlich and Anne H. Ehrlich, *Population; Resources; Environment; Issues in Human Ecology*, 2nd edition, 1972 (W. H. Freeman, San Francisco, USA).
Ralph Nader, *Unsafe at any Speed*, 1972 (Grossman, USA).
R. S. Scorer, *The Clever Moron*, 1977 (Routledge and Kegan Paul, London).
Lewis Mumford, *The Pentagon of Power*, 1971 (Secker and Warburg, London).
Barry Commoner, *The Closing Circle*, 1972 (Cape, London).
Rachel Carson, *Silent Spring*, 1970 (Penguin, London).
E. F. Schumacher, *Small is Beautiful*, 1974 (Abacus, Tunbridge Wells, Kent), *A Guide for the Perplexed*, 1977 (Cape, London).
Dennis Gabor, *Inventing the Future*, 1963; *The Mature Society*, 1972 (Secker and Warburg, London).
Sir George Stapledon, *Human Ecology*, 2nd edition, 1971 (Charles Knight, Tonbridge, Kent).
Kenneth Mellanby, *Pesticides and Pollution*, 1969 (Collins, Glasgow).
M. King-Hubbert, *Energy Resources* (*Resources and Man*), 1969 (W. H. Freeman, San Francisco, USA).
Arthur Koestler, *Janus: A Summing Up*, 1978 (Hutchinson, London).
Robert Jungk, *Brighter Than a Thousand Suns*, 1970 (Harcourt Brace Jovanovich, London).
Robert Ulich, *Progress or Disaster*, 1972 (University of London Press).

Journals such as *Your Environment* and *Undercurrent* are published by people who feel strongly about the future of civilization and ecology.

CHAPTER 2

The way down

2.1 THE FATE OF PREVIOUS CIVILIZATIONS

In the last few thousand years many civilizations have risen to a high level of development and then collapsed. As far as we know none of them had such a sophisticated technology as we have developed, but nevertheless they produced remarkable and long lasting artefacts, such as the great pyramids and the sphinx of Egypt, the arch at Ctesiphon and the extraordinary stone constructions of Middle America. The civilization of Knossos in Crete even had plumbing on the third floor. Some of these civilizations were destroyed by barbarian invasions, others by earthquakes, fires or other disasters and some in Mexico apparently collapsed purely as a result of the irreversible growth of top heavy bureaucracy. In all of them, however, their collapse has had a central cause, which was the run down or degeneration of the overall motivation, spirit, or ethos of the social system. One could describe it as the glue of morality whereby people had sufficient respect for society and caring for other people to be prepared to accept the restraints on their own freedom which were necessary for the maintenance of the stable system. This is the centripetal force that overcomes the centrifugal force of individual selfishness, greed, and lust for power.

The decay of the Roman civilization has been the most carefully studied. The *Encyclopaedia Britannica** says:

> Although during the first century and a half of imperial rule a flourishing local patriotism in some degree filled the place of a wider sentiment, this gradually sank into decay and became a pretext under cover of which the lower classes in the several communities took toll of their wealthier fellow-citizens in the shape of public works, largesses, amusements, etc., until the resources at the disposal of the rich ran dry, the communities themselves in many cases became insolvent, and the inexorable claims of the central Government were satisfied only by the surrender of financial control to an Imperial Commissioner. Then the organs of civic life became atrophied, political interest died out, and the whole burden of administration, as well as that of defence, fell upon the shoulders of the bureau-

* *Encyclopaedia Britannica,* 1958 Edition, Vol. 19, p. 504.

cracy, which proved unequal to the task, and when Rome which was the heart of the Empire lost its rhythm and balance, when Rome no longer had a definite culture, a certain inspiration to impart to the provinces, when Rome's religion succumbed to the several mystical cults brought in by her slaves, when her moral standard yielded before a dozen incongruous traditions, and her literature lost itself in 'blind' gropings after a bygone tradition of a freer day, the provincials in despair abandoned her guidance.

Peter Laurie* in an article about unemployment resulting from the use of micro-processers says:

> We are not the first civilization that has faced this problem; Rome got itself into a fine fix through having an extensive slave system (slave equals micro-processer) and a successful army which achieved such gains abroad that the local populace had nothing to do at home. The usual symptoms appeared, street demos, Coliseum hooliganism.

I think one can clearly distinguish three factors which led to the breakdown of the Roman Empire.

(1) *The destruction of the ideas system,* culture, morality, and religion. This is the glue that holds a civilization together; people willingly accept restrictions on their own freedom because they see that these restrictions are essential to a stable society. A quotation from the *Daily Telegraph* of 10 January 1979 indicates a similar breakdown in our civilization 'As London ambulance men announce that they would join Public Services stoppage, their spokesman declared "That if it means lives lost that is how it must be".'

(2) A rapid growth in the number of bureaucrats writing reports, and more and more replacing real action by committees discussing the desirability of action. This phenomenon is occurring in the Communist countries to an even greater extent than in the Capitalist ones. It is supported by an ever increasing taxation.

(3) *Bread and circuses.* The ordinary Roman citizen was freed from the need to work and given a dole, supported by slaves and tributes from colonies. This caused a total degeneration of self-propelled effort. The slaves are replaced by machines in our age and more and more unemployed are paid a dole and amused with TV and football matches that replace Roman circuses. We in Britain import about half our food, and many people prefer the dole to hard work.

* Peter Laurie, 'About the long goodbye', *New Scientist,* 29 June 1978, p. 929.

Thus, all the signs of the decadence of the Roman Empire can be found in our present civilization.

2.2 PREDICTIONS OF DOOM

In this section I give a selection of quotations from writings of people who have seriously studied the present trends in our civilization due to the application of high technology to perpetual growth, and extrapolated these trends to the future. I do not include the two computer studies organized by the Club of Rome as these are dealt with in more detail in the next section.

2.2.1 *From E. F. Schumacher,* Small is Beautiful, *1974 (Abacus, Tunbridge Wells, Kent) p. 50*

'So the Buddhist economist would insist that a population basing its economic life on non-renewable fuels is living parasitically, on capital instead of income. Such a way of life could have no permanence and could therefore be justified only as a purely temporary expedient. As the world's resources of non-renewable fuels—coal, oil and natural gas—are exceedingly unevenly distributed over the globe and undoubtedly limited in quantity, it is clear that their exploitation at an ever-increasing rate is an act of violence against nature which must almost inevitably lead to violence between men.'

2.2.2 *From Dennis Gabor,* The Mature Society, *1972 (Secker and Warburg, London)*

'Unless wholesale changes are made in our current thinking on education, the role of the individual in society, the organization of, and attitudes towards, our labour force, management and economy, then society will stagnate and collapse under the strain of the potent forces that are rapidly gathering strength. Our present course is set for disaster. . . . If society can cast off its restricting shackles, scrap many of the sacred shibboleths and think instead on severely pragmatic lines, then society may yet be saved. . . . The alternative is death by industrial, atomic or social pollution.'

2.2.3 *From R. S. Scorer,* The Clever Moron, *1977 (Routledge and Kegan Paul, London) pp. 4 and 5*

'We are heading for a period of human misery on a scale quite unprecedented. Throughout the long history of humanity until the industrial revolution, it seems that wealth was accumulated very slowly, and the rich were those who had plundered by conquest. Then Industrial Revolution Man seemed to change all that, for wealth was created rapidly without taking it from another man; or so it seemed. But in reality we were stealing it from posterity. The accounting is about to begin. Man has been prodigal because he has been clever, but his indulgence has made him a Moron.'

2.2.4 *From E. J. Mishan,* Can Britain Survive?, *1971 (Tom Stacey, London) p. 10*

'Only since the last war have men prised open Pandora's box, and discovered technologies for destroying all life on earth many times over and in a variety of ghastly ways. Time, measured only in years, will disseminate this knowledge among smaller and less stable nations that are ruled over by adventurers and tyrants. From this prospect alone one may conclude that the chance of human life surviving the end of the century is not high. To annihilation from military mishap or irresponsibility must be added the possibility of extinction of the species from an uncontrollable epidemic (arising from man's organism being unable to adapt quickly enough to the new and more deadly viruses resulting from wholesale application of 'miracle' drugs), from an ecological calamity (arising from the inadvertent destruction of animal and insect life that preyed on the pests that consume men's harvests while the pests themselves became resistant to the chemical pesticides), and the possibility of a slower and more painful death from choking in the waste products of advancing technology.'

2.2.5 *From Lewis Mumford,* The Pentagon of Power, *1971 (Secker and Warburg, London), quoted in* Anne Chisholm, Philosophers of the Earth, *1972 (Scientific Book Club) p. 7*

'During the last decade, fortunately, there has been a sudden, quite unpredictable awakening to prospects of a total catastrophe. The unrestricted increase of population, the over-exploitation of megatechnical inventions, the inordinate wastages of compulsory consumption, and the consequent deterioration of the environment through wholesale pollution, poisoning, bulldozing, to say nothing of the more irremediable waste products of atomic energy, have at last begun to create the reaction needed to overcome them. This awakening has become planet-wide. The experience of congestion, environmental degradation, and human demoralization now fall within the compass of everyone's daily experience. The extent of the approaching catastrophe and its dire inevitability, unless counter-measures are rapidly taken, has done far more than the vivid prospects of sudden nuclear extinction to bring on a sufficient psychological response. In this respect, the swifter the degradation, the more likely effective measures against it will be sought.'

2.2.6 *From Robert Ulich*, Progress or Disaster?, *1971* (*London Press*) *p. 118*

'Thus we encounter a paradox. On the one hand, modern man is *wealthier, stronger, taller, more athletic, and longer lived than his ancestors. On the other hand he is nervous and sleepless; he cannot meditate or pray.* Even those who should have much reason for contentment cannot find it. And just as internationally the rich and the poor countries become more and more divided from each other, *so also within the same nation the gulf widens between those who have too much and those who have too little.*'

2.2.7 *From Arthur Koestler*, Janus: A Summing Up, *1978* (*Hutchinson, London*) *p. 354*

'The case is this; the contrast between unique technological achievements and an equally unique incompetence in the conduct of its social affairs, have brought the human species to the verge of self-destruction. We who are able to land on the moon and speculate on ultimate mathematical and metaphysical problems have so mismanaged our secular affairs that starvation, race-hatred and nuclear Armageddon may soon write finish to a sordid story.'

2.2.8 *From U. Thant*, People First, *1970* (*Humanist Manifesto*)

'I do not wish to seem over dramatic, but I can only conclude from the information available to me as Secretary-General that the members of the United Nations have perhaps ten years left in which to subordinate their ancient quarrels and launch a global partnership to curb the arms race, to improve the human environment, to defuse the population explosion and to supply the required momentum to development efforts. If such a global partnership is not forged within the next decade, then I very much fear that the problems I have mentioned will have reached such staggering proportions that they will be beyond our capacity to control.'

2.3. COMPUTER MODEL STUDIES SPONSORED BY THE CLUB OF ROME

The Club of Rome was formed by Aurelio Peccei in 1968 to try to think out the major problems facing mankind. Professor Jay Forrester of Massachusetts Institute of Technology had developed a global model for the problems of mankind which could be put on a large computer. These

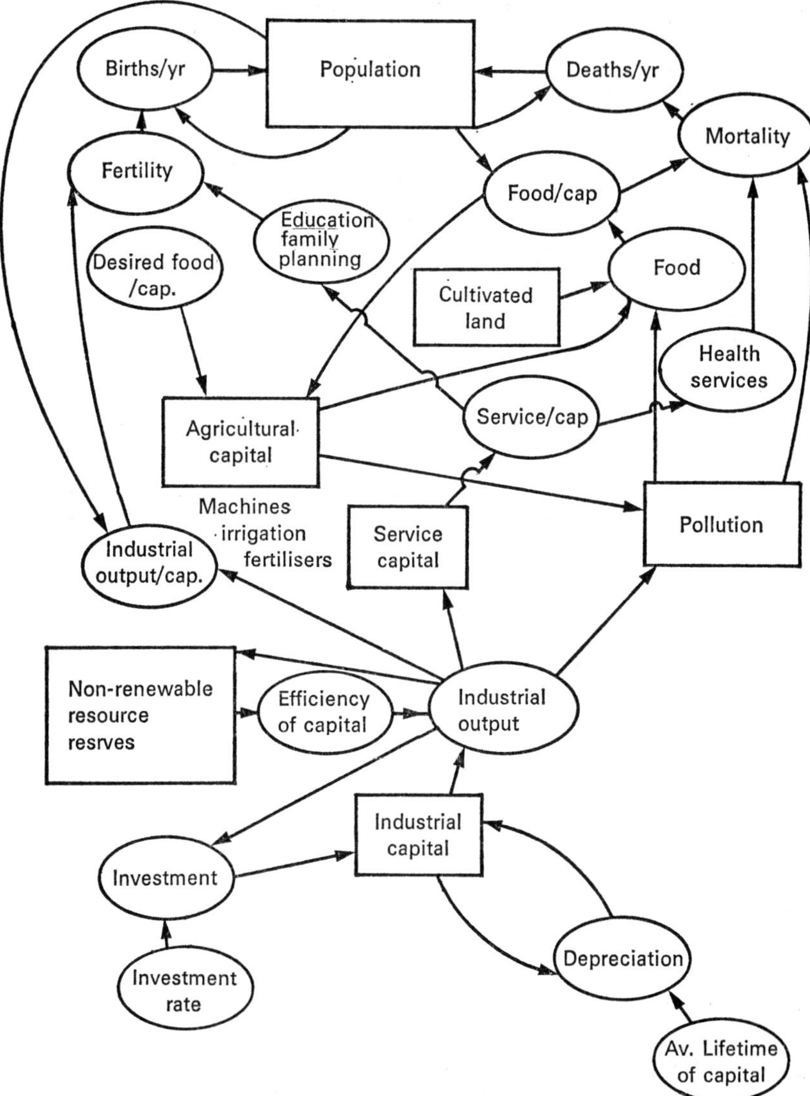

Fig. 2.1. *Feedback loops in the* Limits to Growth *world model* (*Meadows* et al., Limits to Growth, *Figs 24 and 25, pp. 97 and 100, respectively*)

interacting problems are technological and political and the Club of Rome sponsored an international team to examine, with Forrester's type of model, the five basic factors that determined and therefore ultimately limit growth on this planet, population, agricultural production, natural resources, industrial production, and pollution. The result was published in the book *Limits to Growth** Their method was to determine the causal

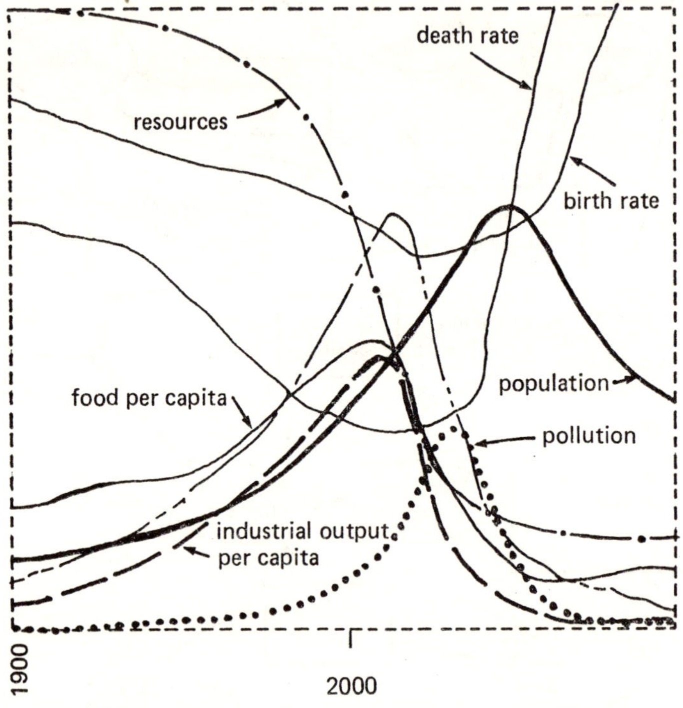

Fig. 2.2. *World model standard run*

relationship between the various input factors by studying the statistics of the first seventy years of this century and then use this causal relationship with information about the limitation of natural resources, predicting what would happen in the future. A simplified diagram of the causal

* D. H. Meadows, D. L. Meadows, J. Randers and W. W. Behrens III, *Limits to Growth*, 1972 (Potomac Associates Books, Earth Island Ltd.).

relations they studied is given in Fig. 2.1, and the resulting world model standard run in Fig. 2.2. Their conclusions were as follows.

(1) If the present growth trends continue unchanged the limits to growth on this planet will be reached within 100 years. The first cause of this is the exhaustion of resources which causes industrial growth to slow right down; somewhat later a massive rise in the death rate occurs due to reduced food and medical services and increased pollution.

(2) It is possible to alter these growth trends and establish ecological and economic stability given certain conditions.

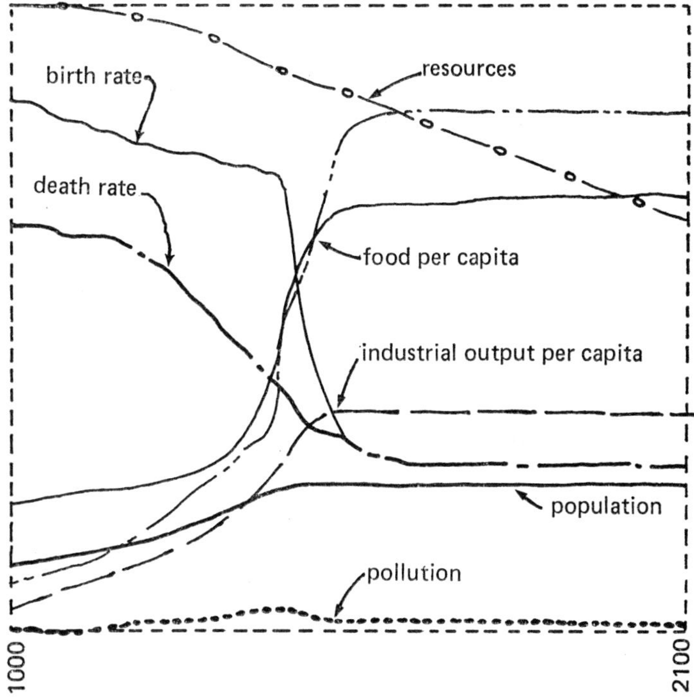

Fig. 2.3. *Stabilized world model 1*

A stabilized world model is shown in Fig. 2.3. This is based on the achievement of the following conditions.

(a) The world population is stabilized by bringing the birth rate down to equal the death rate by 1975*.

(b) Resource consumption per unit of industrial output is reduced to one-quarter of its 1970 value. Industrial capital is stabilized by 1990 by making investment equal to depreciation.

(c) Economic preferences of society are shifted more towards services (e.g., education and health) and away from factory produced goods.

(d) Pollution generation per unit agricultural and industrial output reduced to one-quarter of 1970 value.

(e) To prevent undernourishment of poor, capital is diverted to food production even when uneconomic.

(f) Agricultural capital redirected towards soil enrichment and preservation, e.g., composting urban organic wastes.

(g) To maintain enough industrial capital, design all plant to be more durable and readily repairable.

It was concluded, however, that if all these policy changes were delayed until the year 2000 then an equilibrium state is no longer sustainable and food and resource shortages cause a rising death rate before 2100.

This book was heavily criticized in a number of other books; for example, the World Bank made a special study published in September 1972. The criticisms took two main lines. One was that the report did not take sufficient account of the fact that long before the research limitation occurred man would change the industrial system, and that the other problems of pollution and food could also be solved by similar changes. This criticism is hardly valid, since the whole report was written to persuade power-possessing people to make such changes before it was too late and they would only make them if they were clearly shown what would happen if mankind continued automatically in the present direction.

The second criticism was more valid, namely that they lumped the whole world together, particularly the developed and under-developed countries. This criticism was met by the second computer study for the Club of Rome reported in *Mankind at the Turning Point*†.

* This condition is unlikely to be met before the year 2010.

† Mihajlo Mesarovic and Eduard Pestel, *Mankind at the Turning Point*, 1974 (Hutchinson, London).

In this model the populated world was divided into ten regions, with regions 1 and 2 representing North America and Western Europe, the most highly developed countries, and region 9 South East Asia, the least highly developed.

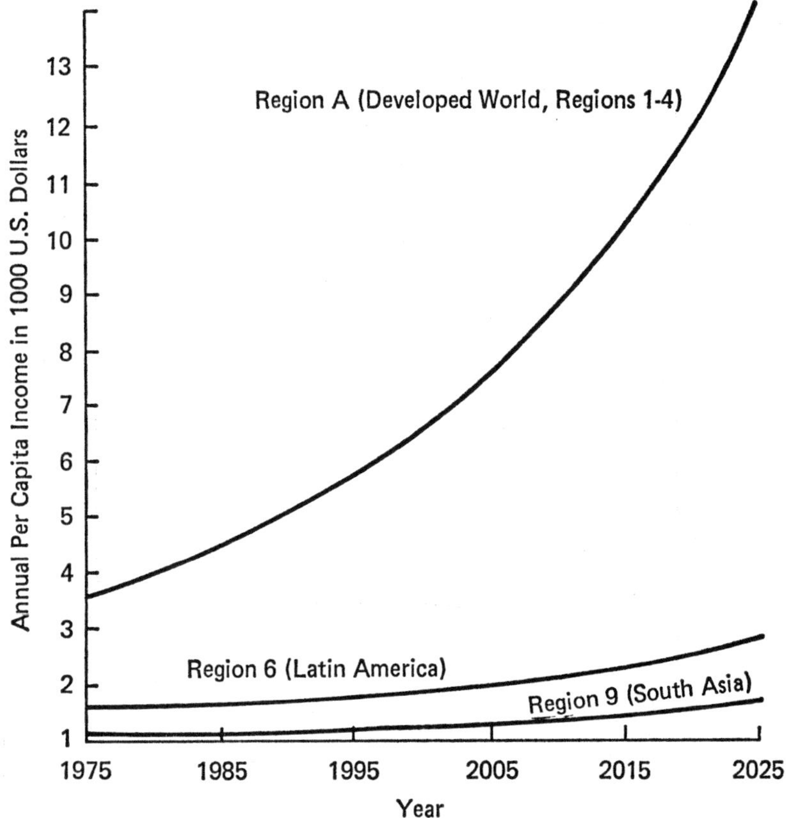

Fig. 2.4. *Interregional gap as projected by historical developments*

Figure 2.4 shows the forward prediction of the widening interregional gap, in per capita income, if everything continues as at present. They suggest present measures of economic development, such as those of highly capital-intensive industry, might worsen the plight of the poor in undeveloped regions by increasing the ranks of unemployed people in large urban areas. What is needed is an 'intermediate technology' which requires capital per job opening about equal to the annual income per employee. This must not require high quality materials, high accuracy, large organizations, or elaborate training.

Figure 2.5 shows various predictions of the future population in the developing world. Curve 1 is a steady continuation of the rate of growth

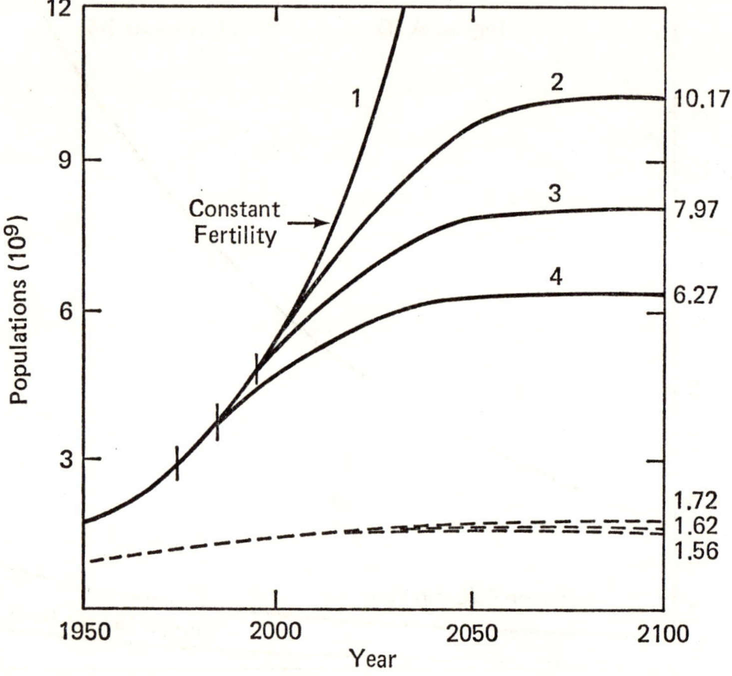

Fig. 2.5. *Population equilibrium for the developing world*

whereby the population of the Third World would amount to 10 000 M people in fifty years. The most hopeful curve, No. 4, is the immediate introduction of a policy with a thirty-five-year transition period to equilibrium fertility. Even this leads eventually to 6300 M, twice the world population at the time of the calculation (1973). The dotted curves represent the increase in population of the developed world in the same period.

In discussing the possibility of producing enough nuclear reactors to provide energy for the whole world on an unrestricted growth scenario they conclude (p. 132) that we would have to build four reactors per week for the next 100 years if they lasted indefinitely. We should then have to have 3000 nuclear parks with eight fast breeder reactors, each producing 40 000 MW, in each. If the reactors last an average of thirty years we should eventually have to build two reactors per day simply to replace those that have worn out. They say: 'the technological fix might well become a Faustian bargain, and worse, because we would not merely be selling our souls, but the well-being and the very existence of generations still unborn'.

They conclude (p. 127) that the problems of the world can only be solved with:

(1) a global approach to the problem,

(2) investment aid rather than commodity aid, except for food,

(3) a balanced economic development for all regions,

(4) an effective population policy,

(5) world wide diversification of industry leading to a truly global economic system.

All of these are essential.

2.4. THE DANGER OF WORLD WAR

Dr Kaplan, director general of 'Pugwash' said:

> But what do we witness today. As Epstein pointed out, between 1960 and 1975 world military expenditures have increased from some 100 000 million dollars per year to three times that amount—which represents a magnitude of waste that is difficult to grasp. This is about three times as much as governments spend on health, about twice as much as they spend on education, and about thirty times as much as the rich countries provide as aid to the poorer countries for their development—all this in the face of hundreds of millions in the poorer countries living on the edge of death from disease, hunger and poverty.*

In an article 'Escalation in the Third World'† Dr Frank Barnaby, director of the Stockholm Peace Research Institute, says that in 1976 total world military expenditure was thirty times what it was in 1900 (at constant prices). It was about forty times the total government economic aid to the Third World, and he says that the world military expenditure exceeded the total income of the poorer half of mankind.

* *Pugwash Newsletter*, October 1976, Vol. 14, Nos. 1 and 2, p. 1.

† Frank Barnaby, 'Escalation in the Third World', *New Scientist*, 21 April 1977, p. 137

An increasing number of Third World countries are developing defence industries of their own to develop and produce weapons.

Arthur Koestler in an article called 'The Brain Explosion'* said:

> Every age had its Cassandras, yet mankind managed to survive their sinister prophecies. However, this comforting reflection is no longer valid, for in no earlier age did a tribe or nation possess the necessary equipment to make this planet unfit for life. . . . We are thus driven to the unfashionable conclusion that the trouble with our species is not an excess of aggression, but an excess capacity for fanatical devotion. Even a cursory glance at history should convince one that individual crimes committed for selfish motives play a quite insignificant part in the human tragedy, compared to the numbers massacred in unselfish loyalty to one's tribe, nation, dynasty, church, or political ideology, *ad majorem gloriam Dei*.

In his book *We All Fall Down*† Robin Clarke says that in the Black Death of 1348–50, more than a quarter of the population of Europe died. He surveys the work being done in chemical and biological warfare and concludes that we have already the knowledge to create destruction of life on a vastly larger scale than the Black Death.

Nigel Calder edited the book *Unless Peace Comes*‡ which describes in horrifying detail the ways in which engineers and applied scientists can extend the ability of people at war to destroy each other and their lands. In this book Philip Noel-Baker, who was awarded the Nobel Prize for Peace in 1959, discusses the way in which the militarists have succeeded for the last 100 years in frustrating all attempts at disarmament and peaceful coexistence. He describes militarists as including:

(1) military staffs who believe that wars are inevitable,

(2) armaments manufacturers and salesmen 'who, encouraged by their national governments to accept patriotic duty, develop a vested interest in preparation for war and in promoting the export of arms',

(3) patriotic societies dedicated to the cultivation of nationalistic ideals and the glorification of military strength,

(4) technical and trade journals with editorials favouring military application of industrial resources,

(5) secret services and intelligence agencies.

Noel-Baker says:

* *Observer Review*, 15 January 1978.

† Robin Clarke, *We All Fall Down*, 1968 (Allen Lane, London).

‡ Nigel Calder (Ed.), *Unless Peace Comes*, 1968 (Allen Lane, London).

Today, militarism assumes subtler guises than those of the bemedalled general gloating over his nuclear weapons, or the mad scientist of the horror film enthusiastically concocting new terrors for mankind; although it cannot be denied that some individuals fit these parts quite well. More typical is the bland civilian who reasons that a disarmed world would be more dangerous, that the 'other side' will agree to disarmament only as a temporary move in the game, that nuclear war may not be as insupportable as is generally supposed.

Novel weapons, from the Maxim gun to the anti-ballistic missile, have created perpetual anxiety among the nations about 'keeping a lead'. The forecasts in this book show that there is no technological plateau on which the contestants can pause for breath, no ultimate weapon with which all will be satisfied. Hitherto, forecasts about new weapons have been largely a prerogative of the militarists, who say to their governments, 'Faster, faster'. May this book serve as an authoritative warning to the majority, and may the cry it evokes be 'Stop'.

It is true that anyone who looks objectively at:

(1) the proportion of scientific and engineering brain power which is being devoted to the improvement of the destructive power of weapons of war,
(2) the proportion of the finance of the rich countries spent on stock-piling weapons,
(3) the enthusiasm of the advanced countries in selling weapons to anyone who can pay for them (often both sides in local conflicts),
(4) the fundamental problem of the jealousy of the poor countries of the extravagance of the rich,

will conclude that we are moving inevitably towards the Third World War.

Another major problem which has to be faced is the danger of proliferation of countries which can make plutonium bombs as a result of having their own nuclear fission power stations. In an article 'The politics of reprocessing'* it is said that times are hard for the nuclear industry because there is a sharp drop in the demand for nuclear power so that the nuclear manufacturing firms are pushing hard for increases of the export market and the development of breeder reactors, but that President Carter objects to the latter for he believes, with good reasons, that they may encourage the proliferation of military weapons.

Once war is started the militarists have complete control and will use any weapons available to win the war as quickly as possible, no matter what the long-term consequences. This is illustrated by the use of fission bombs on Japan and of defoliants in Vietnam. One is forced inevitably to the conclusion that Koestler is right and that a continuation of the

* Frank Barnaby, 'The politics of reprocessing', *New Scientist*, 6 April 1978, p. 18.

present steady growth of armaments and the failure of all attempts at disarmament must lead to the Third World War if we continue, mechanically, in the present direction.

2.5. POLLUTION AND POVERTY

There are rather more than 4000 M people in the world today, of whom about two-thirds live in the less developed countries and have a population increase rate of more than 2 per cent rising in some parts to 3 per cent. Some 70 M people are added every year to the population of this planet. These population growth rates are falling, especially in China where an active policy of the government has been effective, using methods which would not be available in a democracy. In spite of this it is likely that by the year 2000, one in six of the world's inhabitants will be living in India and one in six in China. The reason for this immense expansion of population is largely because the rich civilizations have upset traditional systems of ideas in these countries and have reduced infant mortality, without giving them the increased standard of living and education which has caused the population growth rates in the developed countries to fall almost to an equilibrium level. There is no doubt whatever that this rise in population to well over 6000 M and possibly 8000 M, early in the twenty-first century, will occur and that it will be associated with increasing undernourishment and starvation unless the world food growing problem is solved fairly soon, along the lines discussed in Chapter 6.

Figure 2.6 is a map of the desert areas of the world prepared for the United Nations meeting in Nairobi in 1977*. Pasture land, arable land, and forest land are all being turned into deserts by over-grazing, over-cultivation, by cutting down too many trees, and unwise irrigation. At the present rate we shall lose one-third of the world's arable land to the desert by the end of the century.

Each year we lose, as a result of salination, as much agricultural land as we gain through new irrigation schemes. When there is a dry period the desert breaks out and absorbs good land as a result of what has been done in the better years. The scientific knowledge to turn back the desert is already available but for political and economic reasons it is not being applied.

Mostapha Tolba†, the director of the United Nations Environment Programme, said 'while 40 per cent of the world's population is suffering

* Jon Tinker 'Ancient enemies united against deserts', *New Scientist*, 8 September 1977, p. 582.

† Mostapha Tolba, *The Times*, 3 October 1978.

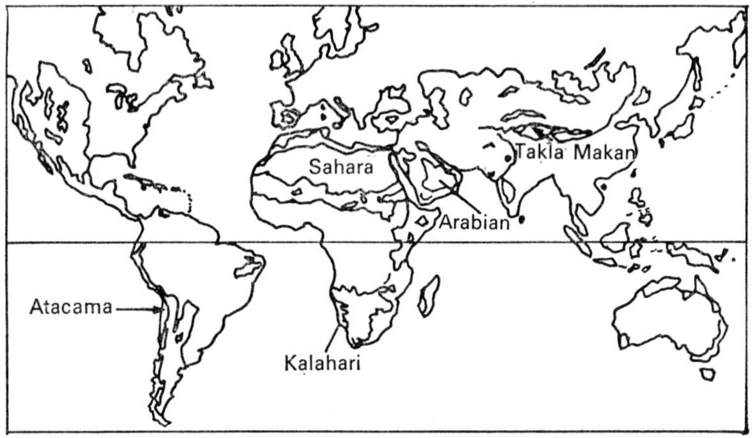

Fig. 2.6. *Desert areas of the world*

from malnutrition we still continue to look upon the agro-industrial residues as waste and pollutants instead of as raw materials for food for humans and animals'. A. K. N. Reddy* said that:

> Science and Technology (S and T) is on trial, accused by the deprived and under privileged section of humanity on three charges.
>
> (1) S and T has ignored the basic needs of the majority of the world's population, and concentrated instead on luxury goods and services for the affluent few, and on military hardware and software to protect this affluence, thereby accentuating the inequalities between and within nations.
>
> (2) S and T has created a skewed demand for skills requiring a few experts with sophisticated expertise for the planning and control of production, and a vast army of workers with the barest minimum of intelligence to execute plans, thereby concentrating power in the hands of experts, and undermining participation and control.
>
> (3) S and T has produced environmental perturbations on a scale and in ways to which nature is totally unaccustomed, so that delicate ecological balances are threatened irreversibly, and perhaps irremediably and catastrophically in the future.

* A. K. N. Reddy, 'Science versus deprived humanity', *New Scientist*, 26 October 1978, p. 270.

2.6. THE PROBLEMS OF THE AFFLUENT SOCIETY

There is much evidence that the short-term political and economic decisions taken in the rich countries, on the basis of giving an immediately rising standard of living to the nation or group, are leading to a falling Quality of Life* even in these countries. R. M. Lala† in an article 'Why we should feel sorry for the rich', suggests that the Baader–Meinhoff gang originated in Germany owing to the break up of family life, but also says: 'A large section of people in the work force of Western nations are ready to withdraw their co-operation from society if they do not get what they want. They demand unprecedented privileges and are not bothered about who picks up the bill.' This, of course, leads to rapid inflation which causes the average standard of living to fall instead of rise. Lala says that there are not enough resources available in the world to meet the rising expectations of the poor nations and the rocketing greed of the rich when both are exercised at the same time.

One can list a series of signs that even the economic system of the affluent societies is breaking down.

(1) Genuine human services (e.g., nursing, police, teaching and public transport) are constantly being whittled away in terms of inadequate staffing, while the paper pushing bureaucracy grows and grows.

(2) The quality of public education is falling due to large classes, dogmatic theories of equality of ability and the possibility of education without pupil effort, and the failure of the educational system to provide a proper education for life. One result of this is rising truantism, another the decay of the self propulsion of the individual.

(3) The degradation of law and order. Sympathy for the violent criminal or rapist exceeds sympathy for the victim. There is a great increase of juvenile delinquency, muggings and other violence by people who would not have any criminal tendencies in a normal society.

(4) Permissiveness. The idea of self-control and self-discipline and hard work to achieve a worthwhile purpose has largely vanished, and the consequences to the individual of promiscuity, drugs, and other body and mind damaging activities are not made clear. These

* 'Quality of Life' is defined in Chapter 3.

† R. M. Lala, 'Why we should feel sorry for the rich', *The Times*, 21 December 1977.

factors are worse in the big towns and big factories than in small ones and in the country.

(5) The farmer and market gardener finds it increasingly difficult to survive economically, as does the small businessman.

There is very clear statistical evidence that in the affluent countries there is a fall in quality of life. This can be divided into three groups.

Firstly, psychosomatic illnesses: coronary thrombosis, stomach ulcers, nervous breakdowns, asthma, even slipped discs, are the most obvious examples. There is even some evidence that the normal life span, having risen to a peak, is beginning to fall again in the richer countries. Joel Greenberg* describes how a town in Pennsylvania had in 1960 a remarkably low death rate with only one-third of the number of deaths from heart attack compared with the rest of the United States. This was when the town still had a traditional old world set of values with very close family relationships. This was upset, and by 1975 they began to get young men dying of heart attacks as frequently as the rest of the USA.

Various forms of cancer are undoubtedly due to the pollutants which get into our food, the air we breathe, and the water we drink, as a result of our carelessness in engineering processes, but there is also a psychosomatic effect in all of these illnesses; that is to say a person who does not feel 'in their heart' emotionally satisfied with their life has much less resistance to illness. They have not sufficient will to live and their body rebels against their life which is unsatisfactory due to lack of both creative self-fulfilment and warm human relationships.

Secondly, when people opt out of society in any way, this again is a fundamental sign that the quality of life in that society is not satisfactory. This opting out may be by suicide, by drug-taking or alcoholism, or simply setting up small communities isolated from the normal economic interactions of society.

Thirdly, vandalism and destructiveness and the growth of violence against individuals are essentially the consequence of the fact that the unsatisfied human needs of a person for creative self-fulfilment and loving human relationships have turned upside down into their own opposites.

2.7. WHERE ARE WE HEADING?

Figure 2.7. is an attempt to express the predicament of our civilization. The irregular circles in the central part of the diagram represent trees in a forest on a hillside. Each tree stands for a problem of humanity,

* Joel Greenberg, 'The Americanization of Rosito', *Science News*, June 1978, **113,** 378.

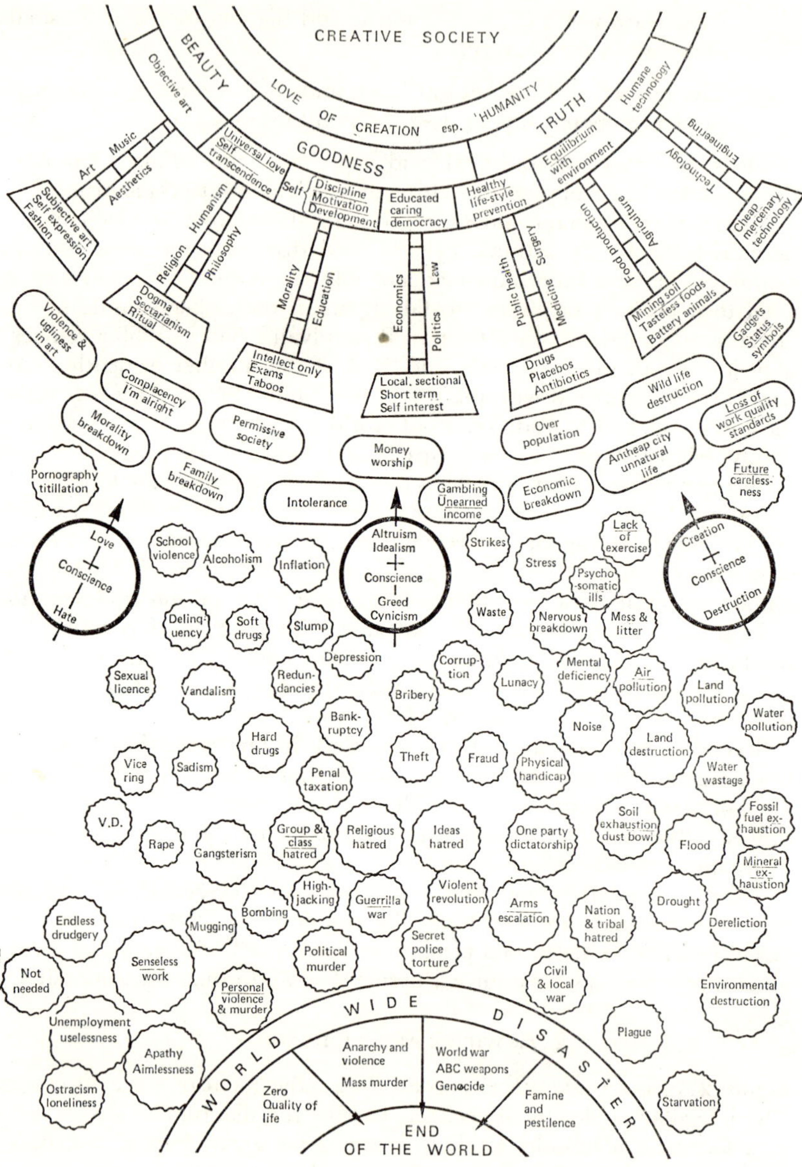

Fig. 2.7. *Either we go up or we go down*

which occurs in many parts of the world. Almost all of them have had world conferences about them which have achieved nothing. When short-term political decisions are taken to avoid one tree the nation immediately runs into a neighbouring tree. This applies to the economic problems in the middle, the psychological problems on the left and the technological problems on the right. Humanity is wandering in the forest in a thick fog. The path of politicians, whose primary aim is to stay in power by avoiding immediate trouble, is a 'drunkard's walk'; as they turn to avoid one tree they run straight into the next.

Such a 'drunkard's walk' must inevitably lead downhill to disaster, since the overall world problems are left unsolved.

This is illustrated by Plate I, where the diagram of Fig. 2.7 is shown in three dimensions. The forest of trees are on a slope so that a random path must lead downhill to the precipice. All the problems represented by the trees on the edge of the precipice are already troublesome in one or more countries.

Four possible disasters which could lead to the end of our civilization are shown at the bottom of the chart. Because technology has shrunk the world these disasters cannot be confined to one country or even one continent, but will inevitably spread over the whole world.

Starting from the right, there is famine and pestilence, killing hundreds of millions of people, which will certainly happen if the under-developed countries continue their population growth unchecked for more than one generation and the rich countries continue to fail to help them to grow their own food.

Secondly, we have the extreme danger of the Third World War with the use of nuclear bombs and chemical and biological weapons. As Koestler points out, this would turn our planet into something as devoid of life as the moon.

Thirdly, a more subtle danger is the gradual erosion of law and order to the point where every citizen is prepared to kill his neighbour in self defence, and roving gangs and war lords destroy the whole system of production and food growing.

Fourthly, we have the most subtle psychological disaster, a system in which human life degenerates to that of the ant heap, so regulated and bureaucratized into a grey uniformity that people do not really care whether they are alive or dead. People have speculated that perhaps hundreds of thousands of years ago the ants were intelligent beings and their civilization degenerated into the rigid machine of an ant heap as their brains withered away and they lost the ability to think for themselves.

Vast concrete blocks of flats are already perilously close to the ant heap in the life they impose on their inhabitants. The architects who design them seldom live there themselves.

A recent article in the *Guardian** describes life in a silicon chip robot age in which the editor of a large newspaper communicates with all the staff scattered over the world by video phone, so that all human relationships, except with one's own family, are indirect, and people never use the skill of their hands. Thus, man's own brain can take him to a totally inhuman world, justifying Scorer's title *The Clever Moron*†.

My own personal conclusion from this study is that it is not yet too late for our civilization to change course and reach the sunny upland where all mankind has the possibility of a life of creative self-fulfilment, but that we have only fifteen or twenty years before it will be too late, and we are irrevocably on the path to one or other of these four disasters. The rest of this book will be devoted to a study of the steps which society in general, and engineers in particular, will have to take to achieve this change of course.

* Peter Large 'The day after tomorrow' *Guardian*, 17 February 1979.
† See Appendix to Chapter 1.

CHAPTER 3

The way up

3.1 THE CREATIVE SOCIETY

I have likened the present position of our technologically based civilization to a group of men wandering in a forest of trees in a thick fog (Fig. 2.7) The forest is on a steep hillside with a precipice at the lowest side. Each time one tree looms up they turn through a right angle to avoid it and promptly find another one facing them. Blind following of short-term selfish objectives causes the resulting 'drunkard's walk' to lead down the slope towards disaster. Carrying the analogy further one can say that to escape disaster they need a map and a compass.

The map shows them that there is a region—a sunny upland or Utopia —where humanity can live in equilibrium with the environment, with all people having the full opportunity of a creative, self-fulfilling life. The compass enables them to take long-term decisions that lead towards this Utopia.

The compass is the human conscience which all normal people possess, although it is often buried in the sub-conscious and overlaid with the self-centred objectives which lead collectively to disaster. The steps necessary to develop and make more readily accessible this conscience I shall discuss in section 3.3. Here I want to explain why I believe that such a machine-served Utopia is at least technically feasible. If this is the case then it is worth struggling towards it and making efforts and sacrifices so that our grandchildren can live in a decent world, a world far better in the human sense than our own, instead of experiencing the total collapse of civilized life.

In the next section (3.2) I shall explain why enormous efforts and sacrifices are necessary if we are to achieve this stability and escape the problems, evils, and dangers of our present situation. We can only do this if we believe there is a real hope of achieving Utopia. I do not mean, of course, that we can ever actually reach such a stable world but we could at least struggle steadily up the hill towards it instead of slipping at an ever increasing rate down towards disaster. I have shown in Chapter 2 that this is what we are now doing.

I call the Utopia, in which all human beings on the earth in the next century could live fully satisfactory lives of self-fulfilment and self-development, to the ultimate limits of each one's possibilities, the *Creative Society*. The Creative Society is shown as the region at the top of the map in Fig. 2.3 and the small plateau reached by the ladders in Plate I. This requires that we have to find a way of giving each one of some 8000 M people all the benefits of the industrial revolution (e.g., excellent diet, homes, education, hygiene, leisure, travel) without any of the disadvantages (e.g., pollution, lack of exercise, strain, unemployment, loss of craftsmanship, fear of violence and war). This in turn requires that we develop only humane technology, with its objective of serving the real human needs of ordinary people, in place of high technology, which enables a very few people to rush about ever faster at the expense of ever increasing energy consumption.

I define the Creative Society as a world society in which all the peoples of the world live in stable or quasi stable* long-term equilibrium with the environment (animal, vegetable, and mineral, destroying as few wild species as possible), and in which every person can find an interesting and worthwhile job which enables them to earn enough to provide for their children and themselves with everything needed for full physical, mental, and emotional health, privacy and companionship, travel and variety, educational development of all their potential capacities and creative self-fulfilment through their own freely chosen artistic or craft skills and hobbies.

In later chapters I shall try to demonstrate that these physical things *can* be provided for 8000 M people but only if the cake of world resources is divided relatively evenly and the rich nations and the rich people of the poor nations are prepared to accept severe curtailment of their present wasteful habits.

I have said that education is a benefit of the Industrial Revolution. However, the education that we have evolved is quite insufficient to lead us to the Creative Society in three respects.

(1) We have developed an intellectual laziness such that the student is not taught to make the tremendous effort required *to think for himself* (*or herself*). Instead we teach dogma, and dogma, however, well meaning when formulated, rapidly becomes disastrous, firstly because the words are almost at once taken in a sense different from that intended, and secondly because the situation soon

* By quasi stable I mean using up fossil fuels and other limited resources as slowly as possible and in a way that leads to an ultimately stable situation without loss of satisfaction of any human requirements. See Chapter 4.

changes. Certainly the only safeguard against dictatorship of all kinds is ordinary people totally refusing to accept dogma or propaganda and insisting on thinking the issues out for themselves. Equally, the only way to reach a stable Creative Society is for enough people to think for themselves about the long-term consequences of our present decisions. We have some educationalists trying to maintain that they can make learning so easy the student has no effort to make at all.

(2) We do not educate the emotional brain, the motivation, or the conscience of the student; indeed the only motivation suggested is 'pass your exams and you will earn more, or at least have a more interesting job'. The achievement of an approximation to the Creative Society, and its maintenance, requires a very thorough education of the conscience and thoughtfulness about what life aims can be truly satisfying to the student.

(3) We suffer badly from Plato's Heresy; the idea that craft and artistic manual skills should be left to the slaves and the gentleman should only occupy himself with philosophy and gymnastics. This often takes the more modern form of the Rutherford Heresy: 'I hope no one finds a use for radioactivity as that would take all the fun out of it'. Everyone should have a balanced education so that they have at least one fully developed manual skill and can lead a balanced life by using this skill as a hobby; the more intellectual they are, the more they need the physical skill to prevent them becoming lop-sided.

3.2. THE NECESSARY CONDITIONS FOR THE SURVIVAL OF OUR CIVILIZATION

The first and most obvious problem relates to the growth of world population. Figure 3.1 illustrates the expected growth of world population from the present level of about 4000 M. It is the general opinion of demographers that this population will certainly grow to 7–8000 M by about the year 2010—that is, in one generation—because of the number of young people already born who will have families. I will try to show in Chapter 4 how we can provide this number of people with enough energy to have all the benefits of the Industrial Revolution and in Chapter 6 how to provide them with a fully adequate diet. However, these studies indicate that if the world population were to double again to 14–16 000 M it would not be possible to provide a stable world system

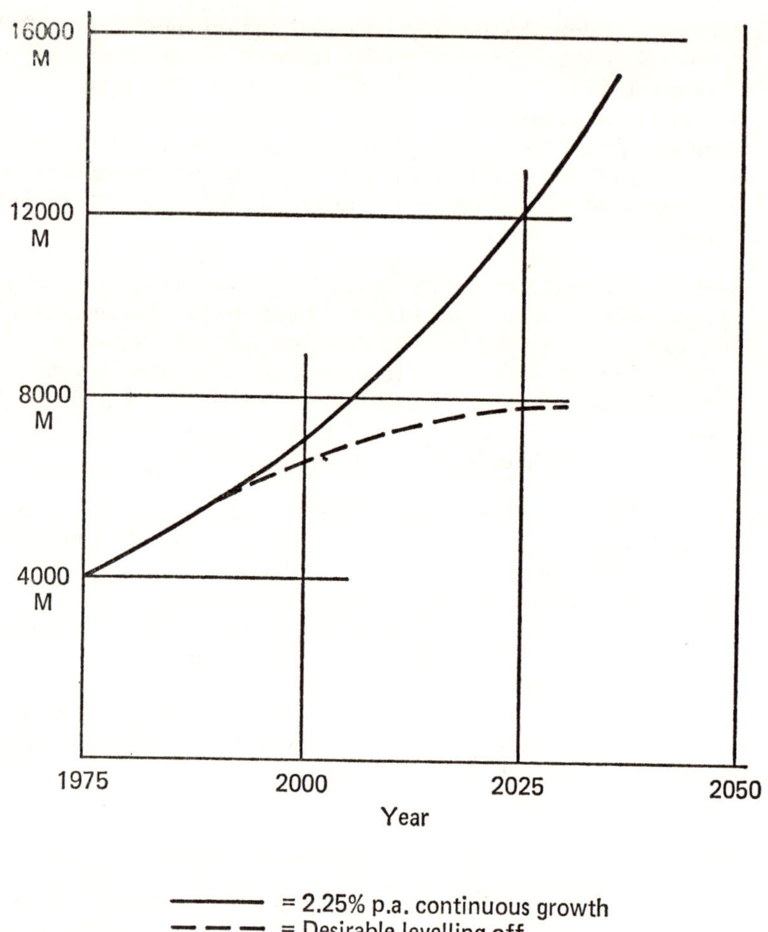

Fig. 3.1. *World population growth*

so that our civilization would necessarily be destroyed by famine and pestilence.

Many writers have pointed out that technology has so shrunk the world that no one country can live in isolation any longer, owing to the interlinking of continents by weapons of war (such as the ICBM), food and raw material transport (especially petroleum), communications, and the spread of disease by aeroplane. Even pollutants, like lead and sulphur, spread from continent to continent. Thus, it is no longer possible for a developed country to say: 'We are all right and it does not matter what happens in another continent'.

If, of course, we have the Third World War within the next generation then the population problem will disappear, but so will our civilization! Thus, the problem we have to consider is that it is essential to find a *humane* way of levelling off the world's population according to the secondary curve of Fig. 3.1. We have to bring the world population growth rate almost to zero within the next thirty years. Now the fact that the population growth rate is already around zero in the developed countries as a result of their good standard of living and education, and that population expansion is occurring almost entirely in the undeveloped countries conclusively demonstrates the following proposition.

Proposition 1. There is only one humane way of levelling off the world's population and this is to provide a fully adequate standard of living and education to all people.

This must be done within one generation if our civilization is to survive. Only when people are educated enough to see the consequences of overpopulation and are able to earn a good living by their own work will they decide voluntarily to limit the size of their families.

The next essential question is how can we avoid the Third World War destroying our civilization within one generation when we are totally unable to stop the continual growth of weapon stockpiles and military expenditure in both the developed and the undeveloped countries. The answer to this question is two-fold. The first necessary condition is that there has to be a fundamental revolution in people's emotional ideas system so that they no longer regard people with another 'football shirt' as non-human characters whom it is legitimate (indeed, a patriotic duty) to kill. This change of ideas system will be discussed later.

The second necessary condition is the elimination of the division between rich and poor countries. As has been said in Chapter 2, so long as poor nations or groups of poor people in one nation look jealously upon neighbouring rich nations or groups who are squandering the earth's limited resources, there will be potentially explosive situations. We can therefore enunciate Proposition 2.

Proposition 2. The enormous differences in standard of living and use of resources between groups of people must be essentially eliminated within one generation if we are to eliminate the tensions leading to the Third World War.

This does not mean, of course, that everybody should have an identical standard of living since people have very different potentials and some

would prefer to make bigger efforts and undergo more training and responsibility to have an above average standard of living. However, the ten-fold discrepancy between the average consumption of energy per head in the developed countries and that in the undeveloped countries will have to be largely eliminated by bringing these two averages roughly to the same level.

Proposition 3. The third proposition derives from the limitation of the earth's resources. If we are to bring the average, per capita, use of resources by all men to about the same level, and there will be twice as many people on the earth as there are now, then *the average, per capita, resource use over all the world must certainly not increase* from the present average. Even that will double the total amount of resources used compared with the present time. Thus, Proposition 3 means that the average per capita use of resources in the developed countries over the next thirty years must decrease to about the present world average per capita figure, and that in the undeveloped countries must come up to this figure. In the developed countries not only do we have to abandon completely the idea of perpetual growth, but we have to find out how to retain all the good consequences of the Industrial Revolution while reducing our consumption of raw materials by several per cent each year for the next thirty to forty years.

Proposition 4. This relates to the poisonous effects of pollution on human life (as studied for example in *The Limits to Growth* discussed in Chapter 2). *No pollutant must be emitted to the atmosphere, to water, or to land until it has been proved conclusively that the level of pollution has no long-term harmful cumulative effect on people, animals, or plants.*

Proposition 5. It is a necessary condition for a stable civilization in the next century that the rich countries gradually eliminate their non-productive activities, such as advertising, weapons manufacture, and fashion and built-in obsolescence, and replace these with genuine attempts to help the poor countries to build up the equipment and knowledge to become fully self-supporting at a good standard of living.

These conclusions have been enunciated by many other people* and are not really open to doubt in their broad sweep, but they do, of course, lead to conclusions which are politically, economically, and socially unacceptable to the rich democracies. However, we can recall Dr Johnson's famous statement that when a man is about to be hanged it concentrates his mind wonderfully. As more and more people grasp the fact that our civilization is moving steadily in the direction of disaster,

* For example J. A. Wagar, 'Growth versus the quality of life', *Science*, 1970, **168**, 1179.

the fact that we have to accept considerable sacrifices and changes of motivation if our grandchildren are to have any life at all will become steadily less unacceptable.

What I propose to outline now are conclusions which I have personally reached*. I hope it will be useful for the engineers to think about these principles from the point of view of the survival of our civilization. I believe that the following five principles are essential to the establishment of a stable world civilization.

(1) A society which is too heavily based on individual self seeking is inevitably self destructive. The centrifugal force of gangsterism exceeds the centripetal glue which is provided by a common acceptance of some rules of morality and altruism, such as 'do as you would be done by'. There is considerable evidence that this is what is happening at the present time in our society, and that our civilization is based far too much on self-interest illustrated by the following facts.

- (a) Politicians become corrupt and concentrate almost all their efforts on preserving their own position.
- (b) The trades unions and other groups tend to pursue their own sectional interests with inadequate consideration of the long-term national consequences.
- (c) Individuals become gangsters or profiteers and are concerned with making money no matter how much others suffer.
- (d) Businesses measure their success only by profits and not by consumer or worker satisfaction.

This principle leads to:

Proposition 6. We have to bring about a fundamental change in the ethos of our society if it is to have any chance of moving into a stable twenty-first century world†.

* None of them are new, indeed they are thousands of years old.

† 'The malaise, I have come more and more to believe, lies in the industrial basis on which our civilization is based. Economic growth and technical achievement, the greatest triumphs of our epoch of history, have shown themselves to be inadequate sources for collective contentment and hope. Material advance, the most profoundly distinguishing attribute of industrial capitalism and socialism alike, has proved unable to satisfy the human spirit. Not only the quest for profit but the cult of efficiency have shown themselves ultimately corrosive for human well-being. A society dominated by the machine process, dependent on factory and office routine, celebrating itself in the act of individual consumption is finally insufficient to retain our loyalty.' Robert L. Heilbroner, 'An inquiry into human prospects', *Observer*, 28 December 1974.

(2) History shows that violence *never* produces a change which is beneficial to the ordinary individual, it merely exchanges one set of tyrants for another; King Stork replaces King Log as in Aesop's fable. It produces immense immediate suffering and long-term bitterness and the new rulers have to use much more violence than their predecessors to keep themselves in power. Violence can only take us away from a better world.

(3) It would only be possible for rulers who have developed themselves beyond the selfish desire for power and become real statesmen to lead a decent society. In view of Lord Acton's dictum 'that absolute power corrupts absolutely', it is clear that only a genuine democracy can be a safe world system. This can only work if both the electors and the elected have been educated, and have developed their own consciences to care for the long-term future, of civilization more than they care for their own short-term immediate advantage.

(4) There must be legal restrictions on consumption and interference with one's neighbours, but these restrictions must be kept to a minimum and do not work successfully unless the general public understands the necessity for them and accepts them emotionally as a necessary basis of society. This is why, for example, prohibition and laws against soft drugs have been almost unenforceable. Many issues such as race relations cannot be put right by law but only by teaching people to care for others.

(5) It is quite essential to a stable society that there is no interference whatsoever with people's thoughts and ideas and all expressions of these which do not cause physical harm to other people. There must be no compulsory dogma; people must be encouraged to think for themselves and work out their own beliefs. There must, however, be an accepted public morality based on an understanding of the harmful effects to oneself and others of inappropriate activities.

3.3 THE HUMAN CONSCIENCE AS A COMPASS POINTING TO A STABLE SOCIETY

In the previous section I have tried to demonstrate the view that the degenerated ethos of our society is essentially incompatible with a stable future for our civilization. There is only one possible way of changing the ethos of society and this is to educate the conscience of the individual so that he or she can develop it to the point where the ethos of caring for the long-term future of humanity becomes much more important than the ethos of measuring success in life by the possession of unnecessary status symbols. They have to realize emotionally that creative self-fulfilment is a much more satisfactory criterion at the end of the day.

Every normal person has a conscience and there are many signs of the action of this conscience, particularly in young prople at the last stages of their education and the first stages of their careers. Earning a sufficient living to provide one's family with a decent standard of living is something totally compatible with the most rigorous conscience. However, it is when one goes beyond this point that worldly ambition for possessions, money, and power begins to overshadow the conscience, and one gets such horrifying beliefs as that of 'the end justifies the means, no matter what they are'. Equally destructive of access to one's conscience is the belief in a rigid dogma.

A normal human conscience imposes duties on its owner which come before any rights. These may perhaps be divided as follows.

3.3.1 *Duties to one's neighbour*

'Do as you would be done by'; in the case of the weak, sick, old, or very young, do as you would wish to be done by if you were in their position. Killing people is wrong under all circumstances.

Being cruel to humans or animals for purely selfish purposes is always wrong. Conversely, helping humans and animals is desirable even at personal sacrifice. At the end of one's life it will be what one gave to others that one remembers with satisfaction, not what one took or received.

3.3.2 *Duties to the world*

One must pay back to society and one's children by personal work and effort at least as much as one has received from one's parents and society (life, health, education, love, friendship). One must leave the environment better than one found it. The Victorian virtues have real validity—saving for a rainy day, standing on your own feet financially and not living on charity or borrowed money.

3.3.3 *Duties to oneself*

Look after your body like an athlete in training in order to be fit to achieve your life's aim. This includes not only obvious things like food, drink, exercise, rest, and hygiene, but also the essential emotional food of perceiving the beauties of nature (trees, flowers, bird song, bird flight, etc.).

However, by far the most important thing which any well developed conscience will tell its possessor is that sooner or later they will die and that it is desirable to judge present actions from the point of view of how they will look when they look back on them from their deathbed. This produces an entirely different set of values expressed in the old adage 'You can't take it with you'. It is from this point of view that I have suggested above that the fundamental ethos of a stable Creative Society would be that people would judge their success in life by the extent to which they had achieved creative self-fulfilment. Put in another way, the criterion would be the extent to which they have used their talents to help in one way or another to make a better or more beautiful world for other people. This would surely seem a more satisfactory judgment of one's life from one's deathbed than the accumulation of status symbols, wealth, or power. The aim is to feel that one has not wasted one's life, so that one will 'perish like a dog', but that one has used one's talents to make something beyond 'self'.

Arthur Koestler, in his article 'The Brain Explosion'*, says: 'Man's deadliest weapon is language. He is as susceptible to being hypnotized by slogans as he is to infectious diseases. And when there is an epidemic, the group-mind takes over.' There is absolutely no doubt that in times of war, revolution or mass hysteria, the fundamental factor which causes individuals to act in ways totally contrary to their conscience is this group suggestibility. In wartime the most successful killer becomes the hero and it is a patriotic duty to kill 'the enemy'.

In the same article Koestler refers to McLean's statement that nature has endowed man essentially with three brains, which he calls reptilian, lower mammalian and the advanced cerebral cortex of the man. G. I. Gurdjieff† locates the conscience in the mammalian or emotional brain, whereas our normal consciousness is in the intellectual brain. This is why it is possible for people to develop, or have suggested to them, ideas systems which cause them to act in ways totally contrary to their true conscience which becomes submerged in the sub-conscious. It was an intellectual ideas system which caused the cruelties of the Spanish Inquisition. It must be concluded that unless, all over the world, the educational system is changed, within the next generation, to one in which people are taught to think for themselves and be totally non-

* Arthur Koestler, The Brain Explosion, *Observer Review*, 15 January 1978.

† G. I. Gurdjieff, *All and Everything*, 1950 (Routledge and Kegan Paul, London).

suggestible to dogma of any kind and to mass hysteria, and their conscience is educated to the point where their natural caring for their fellow men and for life in general becomes a central guiding force in their lives, our civilization will collapse. This is why I have shown the conscience as the compass (in Fig. 2.7) pointing the way out of the forest, a compass which must become available to every thoughtful individual.

3.4. QUALITY OF LIFE AND STANDARD OF LIVING

Clearly from what has been said above, the only hope for the survival of our civilization is for the people living in the rich countries to reduce their consumption of raw materials by a factor of about three over the next thirty years. The only way this can be made politically acceptable is by developing people's conscience to the point where they are prepared to measure their success in life by their creative self-fulfilment, rather than by their possession of status symbols. This concept can be made more precise if we draw a clear distinction between quality of life and standard of living.

Standard of living can be expressed as Gross National Product (GNP) in dollars per capita per annum, or as we shall see in the next chapter in terms of energy consumption per capita per annum. It is a purely materialistic measure of wealth, although it also includes money spent on defence, which is not of benefit to the individual even in time of war, since the enemy will have spent corresponding sums.

Quality of life is a much more subjective measure; it expresses the fundamental emotional feeling of the individual, that his or her life is worthwhile, that they are prepared to face their own death with the feeling that they have not wasted their talents*. As I said in Chapter 2, although quality of life is subjective and not measurable in the laboratory by objective means, there are statistical indices showing degeneration in this quality.

If one plots the daily food input to an adult in kilocalories against the health of the individual, one arrives at a curve as shown in Fig. 3.2. At very low calorie inputs the person is starving; as it is increased further their health rises to an optimum and slopes down the other side as they become obese and 'dig their grave with their teeth'.

Quality of life is subjective but nevertheless more directly real to the individual than quantities objectively measurable in the laboratory or

* H. F. Hrubecky, 'The scientific and technological revolution's demand on education for a future world society', UNESCO Conference, Prague, September 1976, gives a list of six needs for human happiness.

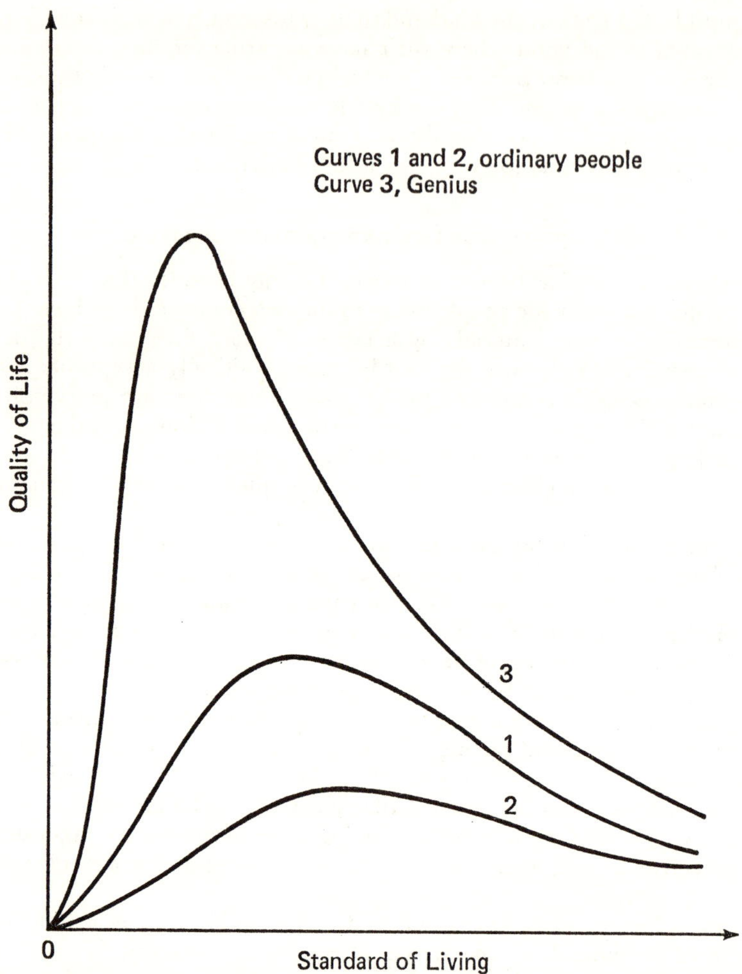

Fig. 3.2. *The relationship between quality of life and standard of living for different people*

the bank. However, we can suppose that it is possible to measure quality of life quantitatively with a scale giving a very high value for a person who is using their talents to the full in creating a better world in their own way (for example, a master painter or composer) or improving the lives of others (a good parent, doctor, or teacher) falling to a zero value for a person who feels their life is not worth living at all. Then it is probable that we could plot quality of life against standard of living and obtain exactly the same relationship shown in Fig. 3.2. At the very lowest standard of living, quality of life is held down by the sheer difficulty of staying alive, it rises to a peak for each individual when they are living at a standard where everything necessary for a satisfactory life is available, and then falls, in an over-rich civilization, down to the extreme of the millionaire whose money is a terrible burden to him and whose friends and wives are more likely to be related to him for his money than for genuine love.

There are many reasons why a very rich civilization has a falling quality of life. People measure their success in life more and more by keeping ahead of the Jones's so that life becomes a worrying rat-race with no time for genuine creative self-fulfilment. Then there is the increasing fear of losing one's possessions by accident, fire, taxation or crime. Finally, there is a nagging guilty conscience at the knowledge that there are vast numbers of people starving in the world. These factors undoubtedly cause a falling quality of life in the societies climbing the scale of affluence.

The undeveloped countries are at an average standard of living below the optimum, while the over-developed affluent societies are above it; hence this diagram (Fig. 3.2) means that the fall in the use of raw materials by the rich societies, which is essential for survival, instead of being a political and social disaster is, if rightly regarded, a blessing, since it can lead to a society in which everyone has far greater opportunity for maximizing their quality of life through creative self-fulfilment than the present one. This is why I am, personally, an optimist and believe it is possible for our civilization to change direction when and if all thoughtful members of the rising generation learn to follow the dictates of their own consciences.

The fundamental basis of a stable world society is a change of public ethos from the motivation of greedy self interest in acquiring more unnecessary status symbols to the judgement of personal success in life by creative self-fulfilment. This applies to all those who have a more than adequate standard of living, whether they are in rich or poor countries because these are the people who have freedom to choose their life objectives. We have to learn to distinguish clearly, and all through life, between standard of living and quality of life.

3.5. THE SEVEN LADDERS

On the chart of the world problems (Fig. 2.7) the lozenge-shaped labels below the bottom of the seven ladders leading up to the Creative Society are the false bases of our present society which makes it fundamentally unstable, so that our civilization is constantly finding itself slipping down the slope among the trees towards disaster. Each of the seven professional ladders has at its base, the *de facto* aims and standards of the corresponding professions, while at its summit I have tried to express the aims and ethics that this profession would have to establish completely if our civilization is to escape from its present unstable situation. The object of this diagram is to explain that the engineer cannot, alone, bring us from our present situation to a world of maximized quality of life, although his contribution is of major importance. The seven ladders thus represent seven aspects of the qualitative change of ethos, which is the sole sufficient condition for survival of our civilization.

Starting with the left hand ladder, it is clear that so long as those people whose profession is concerned with the expression of beauty through art have as their objectives:

(1) self-expression (or even, in some cases, working out their personal obsessions),
(2) making money by leading fashion in a new direction, or by pandering to violence and ugliness,

'artists' are helping to destroy civilization.

In early, more stable civilizations, such as the Babylonian, Egyptian and medieval Christian, art has been an intentional expression of the prevailing ethos, so that it has played a major role in forming the 'glue' which held civilization together. At the top of the ladder *objective art* is taken to mean the art which encourages and helps the deep yearning of man for higher development of his own powers.

In a similar way all the great religions started with a clearly expressed philosophy of man's inner possibilities for self-development to the point where he would transcend his or her own self. They also contained the idea of universal love for all human beings, including one's enemies, and for all created things. These religions have degenerated to the point where these great ideas have disappeared and been replaced by unthinking dogma, which, once it is expressed in words, can be totally misunderstood. They have broken up into endless sects which hate each other and often say 'we are the only ones who will be saved' or 'we have the only truth'. Ritual originally contained inner truths and methods of working on the emotions, but again has largely degenerated into empty formalism. It is quite certain that this world civilization will destroy itself unless the ideas of universal love and the possibility of self development and self transcendence, through many different paths, become widely accepted once again.

Education is the key to the whole possibility of human change. Our present education system largely concentrates on the intellectual brain, but even this is not developed to the point where the student learns to use his intellectual powers for probing deeply into real problems, but rather to the point where he can pass examinations by reproducing half digested facts and other people's ideas systems. The emotional brain, which contains the conscience, as has been pointed out above, is not trained at all, even in the arts subjects. The only motivation given to a student is to conform to the system and to pass examinations to make money. A truly educated emotional brain would mean a person would follow their own conscience, even when it made them unpopular or caused them financial or other suffering. In the old days religious education did attempt to provide some training of the emotional brain, although it was largely in terms of stereotyped dogma. Moral situations can never be expressed in black and white, but have to be assessed by the conscience of the individual in every particular case. Even the physical brain, that is, the creative, artistic, and craft skills and techniques of the body, are largely ignored, especially for those young people who are good at passing exams. An essential condition for the Creative Society is that every person would develop their creative craft skills to the full, as a part of their education, and would exercise them throughout their lives.

The change which is necessary in the group of economics, politics, and law, is quite straight forward. So long as we have local sectional groups pursuing their present short-term aim of a perpetually rising standard of living, our civilization will continue to tear itself apart. We have to have an equilibrium economy, a fully developed standard of living for all humanity, and an educated, caring democracy, since no absolute power can be trusted for long to be concerned with the interests of ordinary people rather than maintaining itself in power. We must have law, sufficient to prevent one man's freedom encroaching too heavily on that of another, but law and government must be kept to an absolute minimum. It is necessary on the one hand to break down, as far as possible, all barriers between nations, races, tribes, religions, and classes, but on the other hand to revert to smallness wherever it is possible in universities, schools, hospitals, policing, and factories.

The situation in regard to public health is equally clear. The medical profession at present concentrates almost all of its efforts on patching people up after they are ill, whether by drugs, placebos, antibiotics or surgery. In a Creative Society almost all the efforts of the medical profession would be concerned with preventive medicine, particularly with ensuring that people knew how to choose a thoroughly healthy and varied diet from which all minute traces of poisons had been excluded, and people knew how to look after their own bodies in regard to exercise, posture, muscle relaxation, freedom from harmful over-indulgence of various kinds, and holidays. Children would be taught how to look after and maintain their body to obtain its optimum working.

Food production will be dealt with in detail in Chapter 6. The main point is that we have to produce enough good food for 7–8000 M people on a permanent, equilibrium basis, not dependent on exhaustible fossil fuels or mined sources of phosphorus and potassium.

Finally, we come to engineering and technology and applied science, which is the primary subject of this book. I shall discuss in detail in Chapter 9 the problems that face the engineer whose conscience tells him that he must replace mercenary technology by humane technology in his own work. By mercenary technology I mean that he is prepared to do whatever he is paid to do regardless of his conscience, which tells him that making machines for killing people or machines which are unsafe or polluting is wrong.

CHAPTER 4

The engineer and energy

4.1. THE PRESENT USE OF ENERGY

4.1.1 *How much is used?*

The Industrial Revolution was based primarily on replacing human and animal muscle energy with fossil fuel energy. Since the total resources of fuels are limited I shall examine in this chapter the strategy that will be needed if the energy is to be available to give the good consequences of the Industrial Revolution to all 8000 M people who will be on the earth's surface in the next century.

In this chapter I shall use the TCE (ton of coal equivalent) as the basic unit because it is large enough so that the annual use per head (TCE/person–year) is expressible in figures between 15 and 0·15 and small enough that one can envisage it as roughly 0·7 m^3 of coal.

Table 4.1 gives a rough equivalence of other units by expressing them in TCE at 100 per cent conversion efficiency. The actual calorific value of coal, natural gas, and oil all vary over quite wide factors, coal particularly because of the variable ash and moisture content. Coal has one advantage over oil in that it contains considerably less of its energy in the form of hydrogen so that the gap between the gross and net calorific values is much smaller*. It is very important to realize the magnitude of these conversion factors; for example, in the case of electricity if one burns the coal in a power station to make electricity and then distributes the electricity and uses it to heat water the overall conversion efficiency is only about 25 per cent† so that one uses 0·5–0·6 kg of coal to produce 1 kWh of heat which will heat 100 kg of water through about 8·6 °C, whereas if one burns the coal directly in a boiler, with 75 per cent efficiency, one needs only 0·2 kg of coal. To produce the same result by hydropower one needs 1 ton of water falling 370 m. Another important comparison is that if a man uses a ton of coal equivalent a year (1 TCE/

* Almost all combustion appliances discharge the combustion products with the water resulting from the combustion of fuel hydrogen in the vapour form so only the net calorific value of the fuel is available.

† The actual immersion heated tank will have 90 per cent heating efficiency; the losses are in the power station and the distribution system.

TABLE 4.1 CONVERSION OF ENERGY UNITS TO TCE (TONS OF COAL EQUIVALENT)

CV (calorific value) of *coal* is 6·0–7·5 Mcal/kg ∴ Mean value 1 TCE = $2{\cdot}8 \times 10^{10}$ J

Natural gas	CV: 35 MJ/m³. So 10^9 m³ natural gas ~ 1·25 MTCE
Oil	CV: 42 MJ/kg or $4{\cdot}2 \times 10^{10}$ J/ton. So 1 ton oil ~ 1·5 ton coal 7·5 bbl oil ~ 1 ton oil. So 5 bbl oil ~ 1 TCE
Electricity	1 kWh = 3600 kJ = $1{\cdot}29 \times 10^{-4}$ TCE (at 100 per cent conversion efficiency). 1 kWh ~ 0·129 kg coal or 1 TCE ~ 7780 kWh. So 1 TCE/year ~ 0·89 kW continuous
Hydro power	1 ton water falling 1 km ~ 2·73 kWh ~ 0·35 kg coal
Heat	1 ton water heated 1°C = 1·16 kWh ~ 0·17 kg coal 100 kg of water heated 30°C (1 hot bath) = 3·5 kWh
Food	3000 kcal/day = $4{\cdot}6 \times 10^9$ J/year ~ 0·16 TCE/year
Nuclear	1 ton uranium total fission ~ 2 MTCE 1 ton uranium in fast reactor (highly enriched) (50 per cent conversion) ~ 1 MTCE 1 ton uranium in thermal reactor (slightly enriched) (½ per cent conversion) ~ 10 000 TCE.

(1) All conversions at 100 per cent efficiency assuming net CV of fuels on weight as sold.
(2) Equivalence sign (~) denotes values assumed with ± 15 per cent accuracy for strategic comparisons.
(3) Gross CV of oil is 7–8 per cent higher than net.
Gross CV of coal is 2–3 per cent higher than net.
Gross CV of natural gas is 10 per cent higher than net.

person–year) this corresponds to about six times his own food energy consumption so that we in Britain, who consume some 5 TCE/person–year, have about thirty energy slaves working for us.

The average consumption of energy per head throughout the whole world's population of 4000 M people is somewhat under 2 TCE*. The extraordinary thing is, however, the immense gap between the use of energy in those countries which have had the Industrial Revolution and those which have not. Roughly speaking 30 per cent of the world's population has an average energy consumption of some 5 TCE/person–year and uses 80 per cent of the total energy of the world. The remaining 70 per cent of the world's population has an average energy consumption of about 0·5 TCE and consumes only 20 per cent of the total energy.

* M. King Hubbert, *Energy and Power*, 1971 (*Scientific American* Book, W. H. Freeman, San Francisco) pp. 31–40.

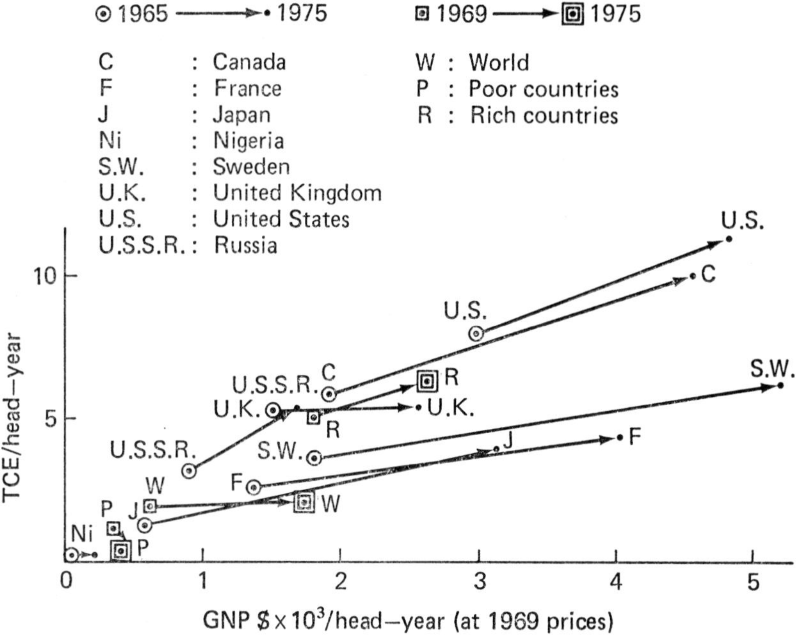

Fig. 4.1. *Wealth and energy consumption*

The most extreme use of energy comes in the USA, where 6 per cent of the world's population use 35 per cent of the total world energy consumption. In Britain we have 1·5 per cent of the world's population and use 3·5 per cent of the energy consumption.

The relation between the wealth of a nation (expressed in GNP/person–year) and the energy consumption (in TCE/person–year) is shown in Fig. 4.1. This shows clearly how there is an almost linear relation between energy slaves and expenditure. Expenditure of course includes money spent on 'defence', the space race, and other factors which are of no direct value to the standard of living of the individual. Nevertheless, it is fairly clear that the countries in which the individual leads a life of relative luxury and wastefulness are those with the highest fuel consumption. In this chapter I shall survey the world's energy resources, both limited fossil fuels and renewable sources, and review the possibilities for fuel economy in the rich countries (that is, achieving effectively the same result for the consumer with far less consumption of limited fuels). I will then establish the principles of a strategy which can lead to a stable world situation in the next century within the limitations of the world's limited energy resources.

Figure 4.1 also shows the changes in the positions for various countries between the years 1969 and 1975. This is reproduced by courtesy of A. Muir-Wood who used it in his address to the Institution of Civil Engineers in 1978. This change covers the period when the price of crude oil went up by a large factor. Thus, countries which came at the upper side of the bands, such as the UK, are those countries which in the earlier period have had traditionally cheap fuel and have obtained less benefit per unit of TCE; that is they have been more wasteful. Those countries which show a change in these six years roughly parallel to the general slope, such as the USA, Canada, and the USSR (indeed the average for all the rich countries) have not shown any improvement in fuel economy as a result of the increases in oil prices. The world as a whole has shown hardly any increase in TCE and a substantial increase in GNP. This is made up of two factors, the rich countries have increased both factors while the poor countries have reduced their energy consumption because they could no longer afford oil for the Green Revolution. At the same time there has been hardly any rise in the real standard of living in the poor countries, so that it is the rise in the standard of living of the rich countries which has caused the average world standard of living to rise.

4.1.2 *The relation of energy to other human problems*

It must be clearly pointed out that the energy problem is inextricably combined with a number of other problems which have to be solved if we are to reach a stable civilization. It is an unfortunate habit of governments all over the world to put each problem in a different ministry: 'the sacred cow of isolation'. This, of course, leads to such nonsensical situations as making consumer goods with built-in obsolescence in order to keep people employed in their manufacture when this means that our descendants will not be able to have these goods at all because we shall have used up all the fossil fuels required to make them. These other problems can be classified under the following four headings.

(1) *The problem of world peace.* It is quite clear to anyone who looks at the relative energy consumption outlined above that it is impossible to achieve a stable, peaceful, world so long as the rapidly increasing population of the underdeveloped countries look at the energy wastage in the developed countries and compare it with their own poverty stricken situation. They have not got the resources to

develop the weapons to attack the developed countries, but the developed countries are busily supplying them with arms and creating unstable situations, in various areas of the world, which are extremely likely, sooner or later, to lead to war between the great powers. Thus it is quite clear that *any energy strategy for the future must take into account the needs of the whole world, and not just those of one country or group of countries.* It is also axiomatic that it must be based on a long-term equilibrium, with ultimately the total use of permanent, renewable energy resources, but with a transition period while the fossil fuels are eked out as long as possible.

(2) *The use of energy is very closely involved in the problems of pollution and noise.* In some cases it is possible to reduce pollution and reduce fuel consumption at the same time by improving combustion to eliminate unburned constituents such as carbon monoxide, soot, and hydrocarbons. However, the American attempt to reduce pollution from cars by the use of catalytic converters and after-burners has had exactly the opposite effect, since combustion has to be worse in order to produce enough unburned material to operate the process. By far the best way of reducing pollution is, of course, to halve the fuel consumption by radical fuel economy methods.

(3) *Consumption of fresh water and production of vapour clouds and thermal pollution.* In Britain electricity generation has been based essentially on the cooling tower and the fresh water evaporated in the cooling towers corresponds to 8–10 tons/person–year*, a figure which would be unacceptable if full account were taken of the value of this water. In other countries the heat is put into local rivers or lakes to such an extent that it seriously damages their natural biology.

(4) *Unemployment.* The example mentioned previously of built-in obsolescence shows one aspect of the relation between fuel economy and unemployment. On the other hand it is equally possible to make fuel economy a means of increasing employment by subsidizing:

(a) those measures of fuel economy which can save much more fuel than is required to install them,

* Hawkins, *Fuel and the Environment*, 1973 (Institute of Fuel). Proceedings of a Conference held November 1973.

(b) devices for using renewable energy sources (e.g., solar and wind power).

Both these are at present uneconomic owing to high capital costs and the cheapness of the fossil fuels.

4.1.3 *What we use energy for*

Figure 4.2 shows an approximate breakdown of the flow of energy through the United States economy prepared by E. Cook in 1971 but revised in 1973*. The units for this table are 10^{15} kcal/year, which is

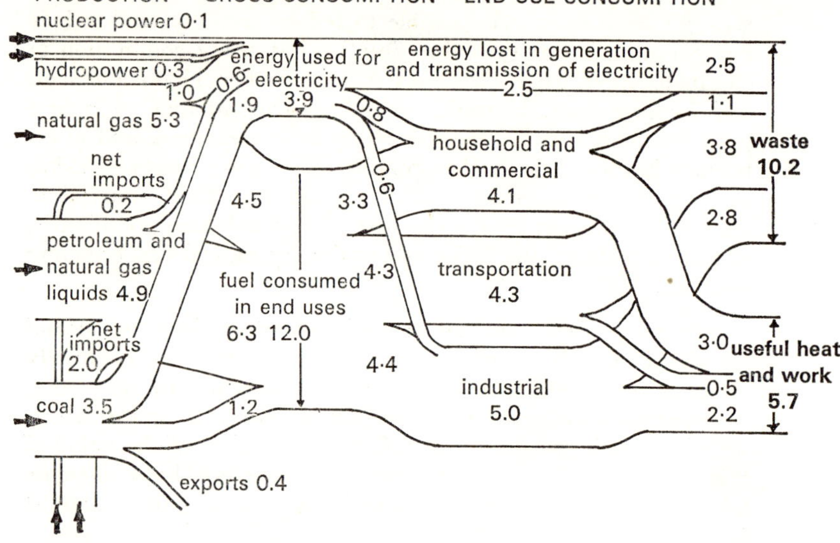

Fig. 4.2. *Approximate flow of energy (10^{15} kcal) through the US economy, 1971*

equivalent to 150 MTCE/year. Thus, the total flow of energy in the USA economy at that time corresponded to 2400 MTCE/year. The corresponding graph for Britain in the year 1972 is shown in Fig. 4.3; here the units are directly MTCE, so at that time the overall flow of energy in Britain was some 360 MTCE, about one-seventh of that in the USA, while the population is about one-quarter, corresponding to our use of about half as much fuel per person/year. These diagrams are based

* E. Cook, *Energy Resources and the Environment*, 1975 (Blackie, Glasgow) p. 33.

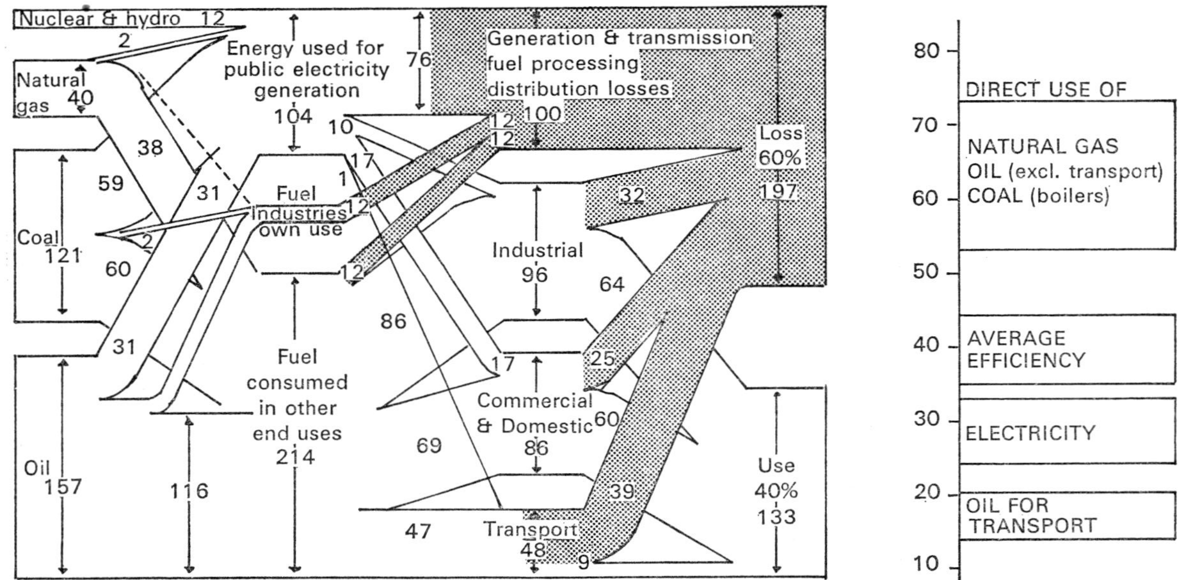

Fig. 4.3. *Approximate flow of energy (MTCE) through the UK economy, 1972*

on incompletely determined but reasonable assessments of the efficiencies of energy use in various industries and other users of energy. The overall efficiency in Britain is given as 40 per cent and that in the USA works out at 36 per cent. It is, of course, very difficult to define what one means by efficiency, and these diagrams are based on rather crude assumptions; for example, when one talks about the efficiency of a car, if one takes a load from point *A* to point *B*, at the same height above sea level, one has, in fact, done no real work; hence efficiency is expressed in terms of work spent against friction of various kinds. This in turn depends on the speed required; for example, canals could take a load of 50 tons pulled by one horse, but very slowly. Figure 4.3* is based on the assumption that the direct use of natural gas, oil, and coal for heating have efficiencies between about 53 and 73 per cent, that the efficiency of electricity generation from fossil or fissile fuels ranges from 24 to 33 per cent, and oil for transport from 14 to 21 per cent. These diagrams show clearly how *the biggest losses are those in the production of electricity and in transport.* Cook assumes electricity generation and distribution efficiency of 31 per cent, transport efficiency of 12 per cent and an overall industrial efficiency of 44 per cent. These figures show clearly how *we could achieve the same results as at present with not much more than half the present fuel consumption, by fuel economy alone.* If we are prepared to be less wasteful in using the products of fuel consumption, as for example, reducing speed of transport, avoiding unnecessary heating and lighting, and eliminating built-in obsolescence, we could probably cut down to about one-quarter of our present energy consumption without any significant drop of real standards of living.

The industrial problem is less clear cut than the transport one, but the three groups of primary interest are the chemical industry, the metallurgical industry, and petroleum refineries and related industries. Between them these three consume more than half the industrial energy. While these industries have all been concerned with their fuel costs for many years, substantial further improvements in efficiency could be obtained at the expense of capital investment. Forty-four per cent of the industrial energy goes to produce process steam and here the efficiencies can come mainly from using less steam rather than improving the efficiencies of the boilers, which are usually already 70–80 per cent.

Turning to the use of energy in homes, offices, and shops, some 56 per cent is used for space heating, 12 per cent in water heating, and 4 per cent in cooking, so that 72 per cent is used for low temperature heating purposes. Much of this comes from electricity, so the figure has

* B. J. Bowden, *Heat. Vent. Engr.*, 1974, **47,** 541, quoted in J. Gibson, 'The Energy Gap', Institute of Fuel, Energy Brake or Break Conference, 1976.

to be multiplied by four to arrive at the fuel consumption in the power station. Lighting as a whole corresponds to only 5 per cent of the total energy consumption and is, of course, a field for which electricity is likely to remain predominant, but even here the total amount used in the USA is 4×10^{11} kWh/year, equal to the flow of energy in the whole of the Alaskan pipeline. Great scope is available for improvement in the efficiency of lighting systems. In the standard filament lamp only a small percentage of all the energy appears as light.

4.2. THE EFFECTS OF MAN'S ENERGY USE ON THE BALANCE OF NATURE*

4.2.1 *Thermal pollution*

Man's extravagant use of fossil fuels produces a number of local climatic effects and would certainly produce more widespread effects if we were to increase our use of fossil fuels to the point where all the 8000 M people who will be on this earth in the next century use as much as the average person in the developed countries at the present time, namely about 5 TCE/person–year. This would involve a tripling of the average energy consumption per head and a six-fold increase in the total world use. Hottel and Howard† point out that the 1970 world energy consumption rate corresponded to 1/6000 that of the solar absorption by the earth, so that direct heating of the overall earth's crust will remain negligible.

One major effect is that of the increase in carbon dioxide in the atmosphere‡. At the beginning of the Industrial Revolution the atmospheric concentration of carbon dioxide was about 290 ppm, but between 1958 and 1968 it increased each year by 0·7 ppm. By 1971 it had reached 323 ppm and it is now increasing at a rate of about 1 ppm/year. This increase in carbon dioxide corresponds to about half the total carbon dioxide given off by combustion of fossil fuels, and represents a rise in the balance point between that emitted and that absorbed by plant photosynthesis on land and in the oceans.

Figure 4.4 shows that there is a large fluctuation each year with the summertime decrease being caused by the photosynthetic uptake during the growing season on land and sea§. It has been predicted that if we continue to increase our fossil fuel consumption at the present rates the figure will reach 380 ppm by the year 2000.

* See also Bert Bolin, *Energy and Climate*, 1975 (Secretariat for Future Studies, Stockholm).

† H. C. Hottel and J. B. Howard, *New Energy Technology*, 1971 (MIT Press, USA).

‡ M. W. Holdgate and L. E. Reed, 'The fate of pollutants', *Fuel and the Environment*, Institute of Fuel Conference, November 1973, p. 113.

§ *Ibid.*, p. 119, Fig. 5.

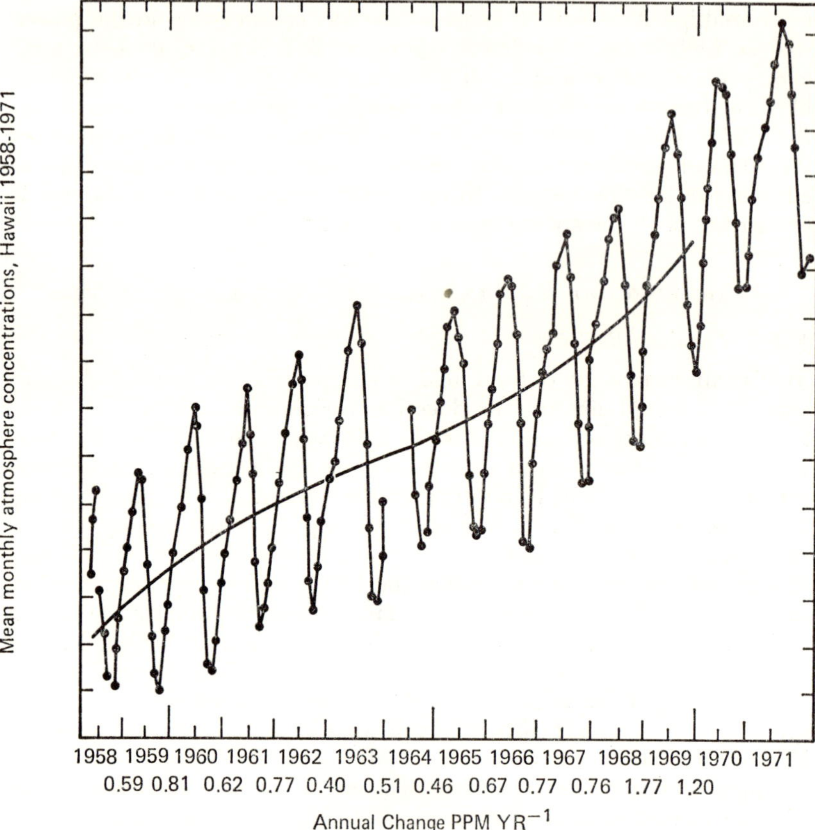

Fig. 4.4. *Mean monthly atmospheric concentration of carbon dioxide, Hawaii, 1958–1971*

Carbon dioxide in the atmosphere causes the 'greenhouse effect' because it is almost completely transparent to solar radiation but has absorption bands in the infra-red which reduce the low temperature radiation of the earth's surface. This means that an increase in carbon dioxide would, by itself, tend to produce a rising temperature on the earth's surface just as the glass of a greenhouse makes it hotter than the surroundings. It has been calculated that the figure of 380 ppm by the year 2000 might increase the mean temperature in the lower atmosphere by about 0·5 °C. On the other hand increasing cloud cover and increasing dust content in the atmosphere reflect more of the sunshine and thus lower the earth's temperature. There is some evidence that at the moment these are more than off-setting the rise of carbon dioxide.

There is some evidence that, in parts of the USSR, the Swiss Alps, and Northern America, the amount of solar radiation reaching the ground has declined by up to 10 per cent over the past fifty years*. This would imply that increased dust emission from man's activities can have a significant local effect.

Another thermal pollution effect arises from the fact that large cities emit a significant amount of heat to the atmosphere and can raise the air temperature locally. When this energy release is more than 10 W/m^2 (i.e., above 10 per cent of the insolation) there is a warm air layer above the city held in by inversion, with increased haze and fog, which traps pollution (see Fig. 4.5).

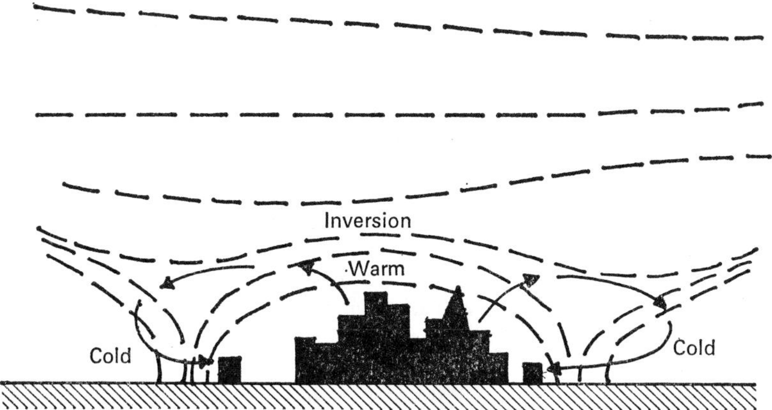

Fig. 4.5. *Local air circulation created by heating within a city (from Bert Bolin,* Energy and Climate, *1975 (Secretariat for Future Studies, Stockholm)*

Finally, there is the heat emitted by large thermal power stations which are not combined with the use of pass-out steam for total energy. These stations have to exhaust some 45–55 per cent of the energy of the fuel consumed from the turbine at as low a temperature as possible to maximize the overall thermal efficiency. In Britain we have concentrated

* M. W. Holdgate and L. E. Reed, 'The fate of pollutants', *Fuel and the Environment*, Institute of Fuel Conference, November 1973, p. 119.

mainly on the large cooling tower in which the water which has cooled the condensers is trickled over a distribution stack at the base of an enormous natural draft chimney. A significant amount of this water evaporates so that a cooling tower serving a 2000 MW power station evaporates 5000 m^3 of water a day*. This means that more than 8 tons of river water are evaporated for every person in Britain every year. Although these stacks are very high, there are cases where this extra moisture brings the ambient air above saturation and creates a thick fog. There used to be a problem with droplets being carried up the towers which led to water deposition on nearby roads but this has now been solved by droplet eliminators. A dry cooling tower system has been installed, but the capital cost is a little higher and the overall thermal efficiency a little less.

In the USA they have tended to put this waste heat mainly into lakes and rivers, and this has, in some cases, raised the temperature so much as to upset the normal marine life. In cold climates it is, of course, possible to cause an improvement. However, there is evidence that really high energy releases (30 kW/m^2) can generate tornadoes, so there is an upper limit to the discharge from power stations!

Bert Bolin has made a careful study of the results of a ten-fold increase in mankind's total energy consumption. He concludes that the regional effect might amount to a 1°C temperature rise at most, but that the global changes represent a serious problem. The direct heating would only increase the mean temperature by 0·1°C, but a doubling of the carbon dioxide in the atmosphere would imply changes in the energy flux—on a global scale about 1 per cent of natural ones—which would noticeably disturb the present radiation balance and could increase the temperature by 2 °C†. If we increase our global consumption of fossil fuels by 5 per cent/year the carbon dioxide content of the atmosphere would double in 50–60 years, and if the chemical equilibrium of carbon dioxide from atmosphere to sea (the principal sink) is upset, this doubling time could be considerably less. When we remember that the average per capita energy consumption of the 70 per cent of the world's population in undeveloped countries is one-tenth that in the developed ones, it is clear that it is certainly not possible for the whole future world's population to use as much as the developed countries use now.

It has been suggested‡ that the rise in carbon dioxide in the atmosphere

* Hawkins, 'The electricity supply industry', *Fuel and the Environment*, Institute of Fuel Conference, November 1973, p. 69.

† J. Gribbin, 'Fossil fuels: future shock', *New Scientist*, 24 August 1978, p. 541.

‡ R. Lewin, 'Why the yule logs must not burn', *New Scientist*, 23 December 1976, p. 750.

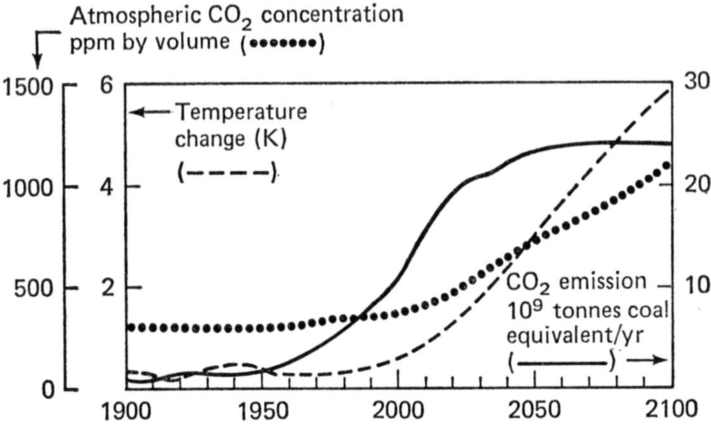

Fig. 4.6. *Strategy for carbon dioxide impact of 30 TW fossil fuel*

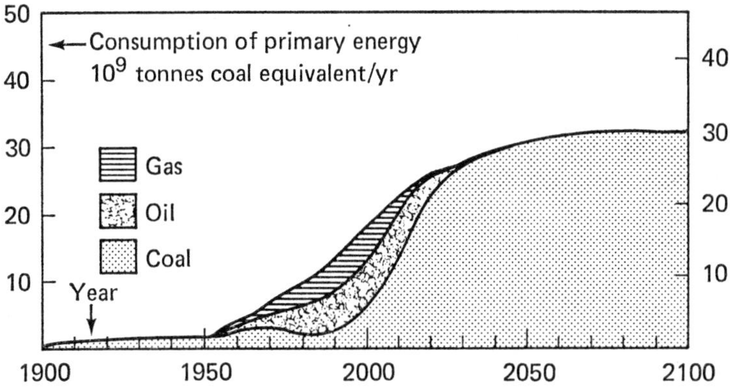

Fig. 4.7. *Future energy production from fossil fuel*

may be due as much to the destruction of the world's forests resulting from the expansion of human population as to the 5 Gt (1 Gt = 10^9 tons) of fossil carbon burnt every year. The amount in the atmosphere increases by 2·5 Gt tons each year and the rest goes into the sea or land biomass. The total carbon in the atmosphere is about 700 Gt while the land biomass contains 1800 Gt, the deep ocean 32 000 Gt.

Gribbin concludes from the National Academy of Science report

Energy and Climate, that an increase of world fossil fuel consumption to 30 TW (1 TW = 1130 MTCE/year) would produce carbon dioxide and temperature increases as shown in Fig. 4.6. The energy would all be produced from coal after 2050 when the oil and natural gas were exhausted as shown in Fig. 4.7.

Marland and Rotty* consider a model in which:

(1) USA energy growth is from 73 Q in 1974 (1 Q = 10^{15} BTU/year = $1{\cdot}055 \times 10^{18}$ J/year = 38 MTCE/year) to 125 Q in 2025, 15 per cent being non-fossil energy (mainly nuclear and hydro),

(2) Western Europe, Australia and Japan have 2 per cent/year growth in total energy, 15 per cent non-fossil,

(3) centrally planned European economies have 4 per cent/year growth, 25 per cent non-fossil,

(4) centrally planned Asian economies have 5·1 per cent/year growth, all fossil,

(5) developing countries have 5·5 per cent/year growth, all fossil.

This leads to a five-fold increase in world energy consumption from 240 Q in 1974 to 1250 Q in 2025; most of this would be from coal, and about half the carbon would remain in the atmosphere.

4.2.2 *Pollution due to incomplete combustion*

In general, pollution due to carbon monoxide, unburned soot, and unburned hydrocarbons can best be eliminated by improving combustion so that fuel economy and pollution elimination here go hand in hand. Most of the man-generated carbon monoxide comes from petrol engine motor vehicles and it has been estimated† that the total amount man emits is considerably smaller than that emitted from the surface of the oceans from the oxidation of methane and from the degradation of chlorophyl on land. Carbon monoxide is probably converted to carbon dioxide in the stratosphere and on suitable soils. However, carbon monoxide can be very serious in urban areas, especially in road tunnels and confined spaces with heavy traffic. Carbon monoxide is, of course, poisonous in very small concentrations‡ and causes headaches at even lower concentrations. Petrol engines are adjusted to run on the rich side

* G. Marland and R. M. Rotty, 'The question mark over coal, pollution, politics and CO_2', *Futures*, February 1978, p. 21.

† M. W. Holdgate and L. E. Reed, 'The fate of pollutants', *Fuel and the Environment*, Institute of Fuel Conference, November 1973, p. 118.

‡ Continuous exposure to 100 ppm carbon monoxide creates serious physiological problems (W. S. Broecker, 'Man's oxygen reserves', *Science*, 1970, **169**, 1537).

of stoichiometric to give a better response to sudden acceleration. Exhaust gases would be dispersed much better in most situations if the exhaust was emitted at the roof level of the vehicle instead of close to the ground where the air is much more stagnant and tends to be picked up by the vehicle behind in a traffic jam.

Diesel engines run with excess air under all conditions, except where the engine is being asked to produce significantly more than its rated maximum power, since they operate with a more or less constant air intake and control the power by the amount of fuel injected. Heavy smoke from a diesel comes either from this cause or from bad maintenance.

Britain's Clean Air Act in 1955 was introduced largely because of the very smoky combustion used in domestic open fires and in many domestic boilers and furnaces burning high volatile bituminous coal. The Act largely resulted in the movement away from coal for these purposes to the use of town's gas and more recently natural gas. When these fuels become no longer cheaply available it will be possible to go back to coal and burn it smokelessly for all these purposes with the application of existing knowledge.

4.2.3 *Sulphur*

Most coals and residual oils (the heavy fraction left after refining for petrol, kerosene, etc.) contain 1–3 per cent sulphur. In normal combustion this appears in the stack as sulphur dioxide, but if there is a lot of excess oxygen some of it can be oxidized to sulphur trioxide resulting in problems with the deposition of sulphuric acid in the cooler parts of the systems. The sulphur dioxide emitted into the atmosphere is eventually oxidized to sulphur trioxide and washed out as sulphuric acid. There is no evidence that sulphur dioxide is accumulating in the atmosphere* and it has been shown that its lifetime in the atmosphere could vary from one hour to a few weeks. In Britain so far the policy has been for the large power stations to have very tall stacks so that at no point is the ground level concentration very high. Bankside power station had wet cleaners, but of course this reduced the gas temperature and therefore brought the peak ground concentration point much closer to the base of the stack. The total UK emission of sulphur at the present time is estimated to be 6 M tons/year. It appears that the sulphuric acid does no harm, provided it is widely dispersed, in a country like Britain, which has principally limey soil, but that it can be very harmful to trees and other plant and fish life in countries like Norway and Sweden, which have essentially a

* M. W. Holdgate and L. E. Reed, *ibid.*, p. 120.

granite base and therefore an acid soil. There is evidence* that sufficient of the sulphuric acid from the British and other Western european power stations reaches Norway and Sweden to be a substantial part of the cause of damage there. It is suggested that this occurs as a result of an adsorption of the gas on fine particles so that it is not all washed out as it is carried across by the prevailing winds. Sulphur dioxide and sulphur trioxide can be extracted from combustion gases by both wet and dry processes, but these are expensive and require the handling of large amounts of material. Hottel and Howard quote figures of around 0·0025 $/kWh and regard the desulphurizing of the fuel in a pre-combustion stage as more attractive.

4.2.4 *Oxides of nitrogen*

High temperature combustion, especially in the presence of excess oxygen can produce significant amounts of oxides of nitrogen. In general natural sources emit some thirty times as much as man's activities and in any case, oxides of nitrogen can finish up as a useful source of fixed nitrogen for agricultural purposes. However, they cause a real problem in the Los Angeles area because there they have a combination of week-long thermal inversions, the emission of a great deal of hydrocarbons, and high solar irradiation. These result in the formation of ozone which is harmful to humans (particularly causing eye irritation), to vegetation, and to rubber, causing it to perish. Citrus growing is said to be uneconomic within some 80 km of Los Angeles†.

4.2.5 *Lead*

Lead is an insidious poison which accumulates in the central nervous system causing irritability and eventually, in higher concentrations, madness and death. It is possible to trace the increase of the lead in the snow in Greenland because a new layer is laid down each year†; the results are shown in Fig. 4.8. The first rise corresponded to the increase in emissions after 1750 resulting from the beginning of the Industrial Revolution with the great increase in the lead smelted. Since 1940 it has been rising rapidly due to the use of lead as an additive for petrol (400 ppm in motor gasoline). Some 350 000 tons was used in 1970 in the world as a whole, particularly in the northern hemisphere where most of the cars are. The UK used about 10 000 tons in 1971. Approximately 70 per cent of this lead is emitted as an aerosol from the exhaust pipe, probably attached to a small aggregate of carbon particles, the whole particle having a mean size less than 1 μm. The dust in most urban streets

* R. J. Crookes, 'Environmental pollution', *Energy and Humanity*, 1974 (Peter Peregrinus, Stevenage) p. 46. Also M. W. Holdgate and L. E. Reed, *ibid.*, p. 117.

† M. W. Holdgate and L. E. Reed, *ibid.*, p. 120.

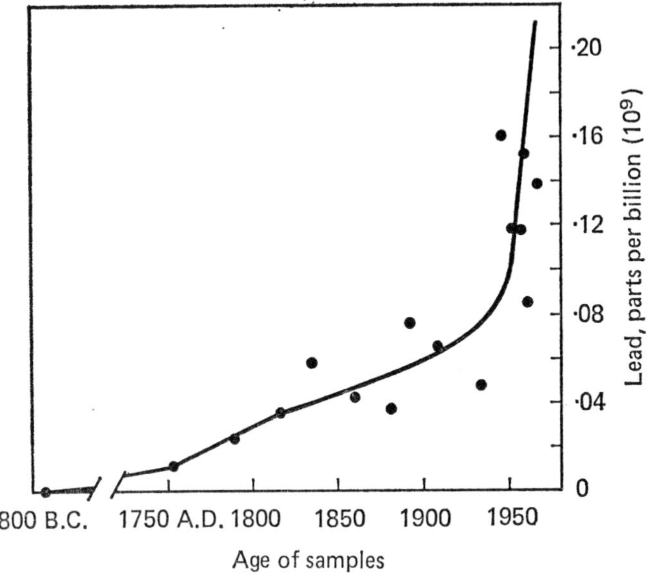

Fig. 4.8. *Increase of lead in snow at Camp Century, Greenland, since 800* BC

carrying a reasonable amount of traffic contains 0·1 to 0·3 per cent* but concentrations ten times higher may occur around some industrial sources. The atmosphere in Fleet Street has 3·2 $\mu g/m^3$†. It can be shown that the amount of lead around a motorway falls off rapidly as one goes away from it. A recent study‡ shows that sheep grazing near busy roads have so much lead in their kidneys, liver, and lungs that the consumption of this part of the animal could be seriously dangerous for humans. A very simple, already existing possibility, for eliminating lead completely from vehicles, and saving fuel at the same time, would be to change from spark ignition engines to diesel.

4.2.6 *Miscellaneous pollution effects from combustion and the use of fossil fuels*

Hydrogen fluoride and hydrogen chloride are emitted in significant quantities from certain industrial systems§ and can cause damage to

* M. W. Holdgate and L. E. Reed, *ibid.*, p. 120.

† *Air Pollution from Road Vehicles*, National Society for Clean Air, Report, 1967.

‡ 'The road to lead in sheep', *New Scientist*, 28 September 1978, p. 921.

§ See, for example, *Industrial Air Pollution*, 1975 (Health and Safety Executive, London).

vegetation, and animals grazing on it, in the neighbourhood of the stacks.

Pulverized fuel combustion systems have efficient gas cleaning to take the ash out and it is now possible to make effective use of this material as light-weight concrete aggregate for land fill. All solid fuel combustion systems require fully adequate gas cleaning, ranging from cyclones for small systems to electrostatic precipitation, bag filters, and wet washers for large ones.

There has been much complaint about the ugliness of power lines with great pylons stalking across beautiful valleys. This appears to be an inevitable consequence of an overall grid system for electricity since carrying the cables underground would cost at least sixteen times as much as overhead construction. This is one of the reasons why the installation of large scale power stations all over the world to provide electricity for 8000 M people is not a feasible solution to man's problems.

Another problem which must be mentioned is the discharge and spillage of oil from large tankers at sea. The oil forms a thick tarry emulsion with the sea water which dirties beaches all over the world and causes much damage to marine life and to sea birds. It is known how to prevent this completely, but it is a matter of spending more money.

4.2.7 *Noise*

By far the most important combustion noise is that from vehicles, especially large lorries and aircraft, but there are also serious problems with industrial noise. Turbulent flames can cause acute noise emission especially from high-intensity oxy-fuel burners. Fluctuation of the flame front can be noisy, and anchoring it eliminates this effect. Research can reduce these problems very much.

4.2.8 *Conclusion about pollution from fossil fuel combustion*

(1) The changes in the weather caused by increased carbon dioxide in the atmosphere make it essential to restrict the overall world use of coal and wood as fuels to not much more than twice the present figure.

(2) Much more money must be spent on reducing all the other pollutants and noise emitted to the atmosphere by combustion of fuels. This requires both further research and capital investment in known methods.

4.3. FOSSIL FUELS

4.3.1 *Coal*

The solid fuels—anthracite, bituminous coal, brown coal, and peat—are by far the largest reserves of fossil fuel in the world. They range from thick layers near the surface, which can be mined by opencast methods, to deep and thin seams under land and sea. They are present in much greater quantities than natural gas and liquid fuel because the process of accumulating thick layers of organic matter, which was then covered by sedimentary rocks, was much more widespread in nature than the preservation of liquids and gases from organic matter impregnated in permeable rock beneath inverted saucers or domes of impermeable rock, and held in place by water underneath. All three kinds of fossil fuel often have high sulphur content. In the case of solid fuel the sulphur may be in the ash as pyrites or in the combustible material as organic sulphur compounds. In either case it tends to finish up as oxides of sulphur in the combustible gases.

Coal has been mined since Roman times when this was regarded as the worst fate to which a slave could be subjected. Working in a mine is still dangerous, in spite of immense improvements in safety precautions. It is also necessary to make long dangerous underground journeys, daily, from the shaft to the coal face, and the increase of mechanization has increased the difficulty of ventilating well enough to keep the dust content sufficiently low to prevent face workers from suffering from pneumonoconiosis at the end of their lives. Modern developments in coal cutting machinery, such as the ranging drum shearer, and in coal conveying systems, such as the armoured chain conveyor, have enormously increased the output per man shift, but they still have frequent breakdowns which cause total loss of output for significant times, and, in spite of the fact that the 'longwall' method has replaced the old 'room and pillar' method, the total fraction of the coal won is still only about 40 per cent.

In spite of these disadvantages the Industrial Revolution was essentially based on the expanding use of coal up to the First World War, at which time Britain was mining 270 M tons/year. At that time, as shown in Fig. 4.9*, coal was providing about 75 per cent of all the energy in the USA, but the percentage has been decreasing there ever since, mainly from the increase in the use of petroleum and natural gas. Figure 4.10 shows the production of coal in Britain up to the year 1970 and Fig. 4.11 shows the way in which coal has actually been displaced by oil in Britain since then.

A survey of the position in other countries shows that, in some of them, coal output has been increasing during the period that British output

* From H. C. Hottel and J. B. Howard, *New Energy Technology*, 1971 (Massachusetts Institute of Technology).

Conversion factors
10^{10} BBLS
crude oil
nat gasoline
LPG
10^{13} KWH (thermal)
10^{9} tons coal
10^{9} cords wood 1830–85 1905 +
10^{13} cu ft nat gas
10^{12} kWh (32% eff 1968)
total energy
oil
coal
Population ordinate 2367
natural gas
wood
hydro power
anth & bit coal
crude oil & nat gas liquids
nuclear power
Quadrillions (10^{15}) BTUS and Population × 10^{-7}
100
80
60
40
20
10
8
6
4
2
1
0·8
0·6
0·4
0·2
0·1
'50
'60
'70
'80
'90
1900
'10
'20
'30
'40
'50
'60
'70
'80
'90
2000
Year
100
10
1

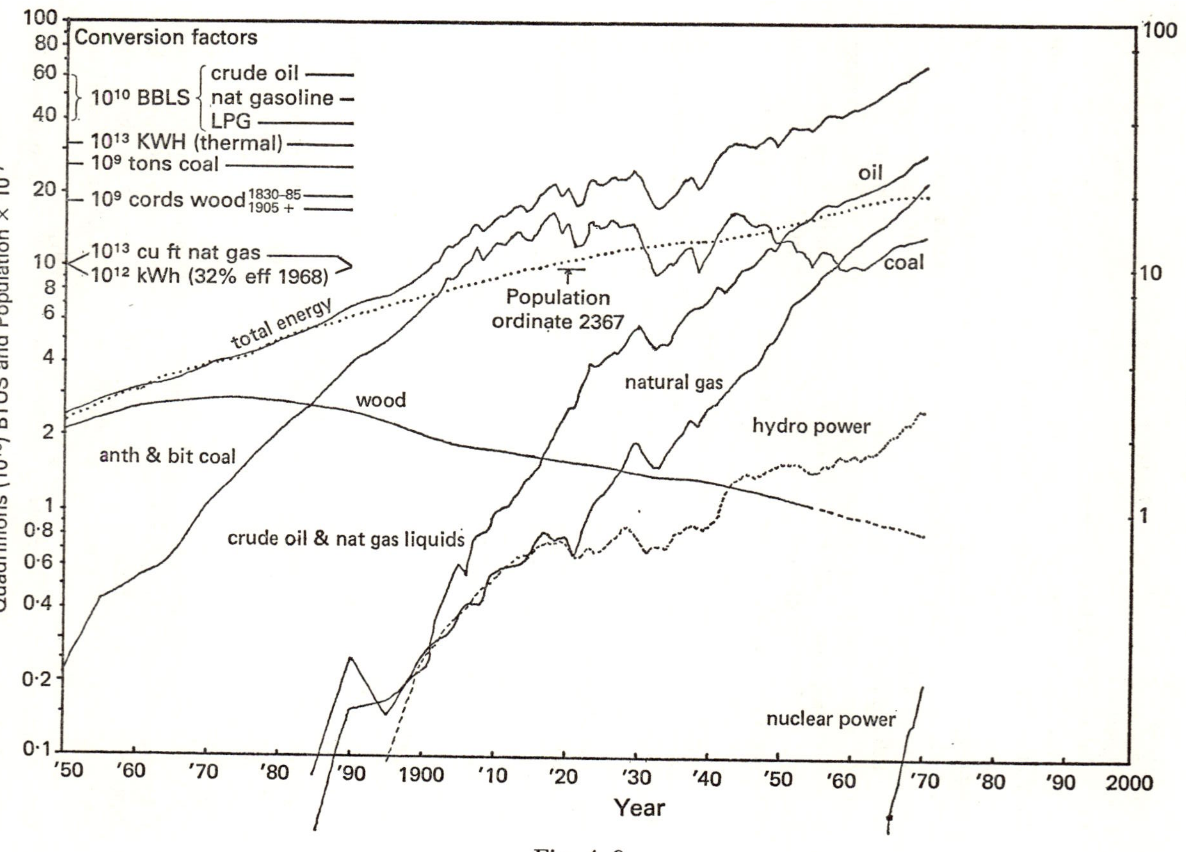

Fig. 4.9

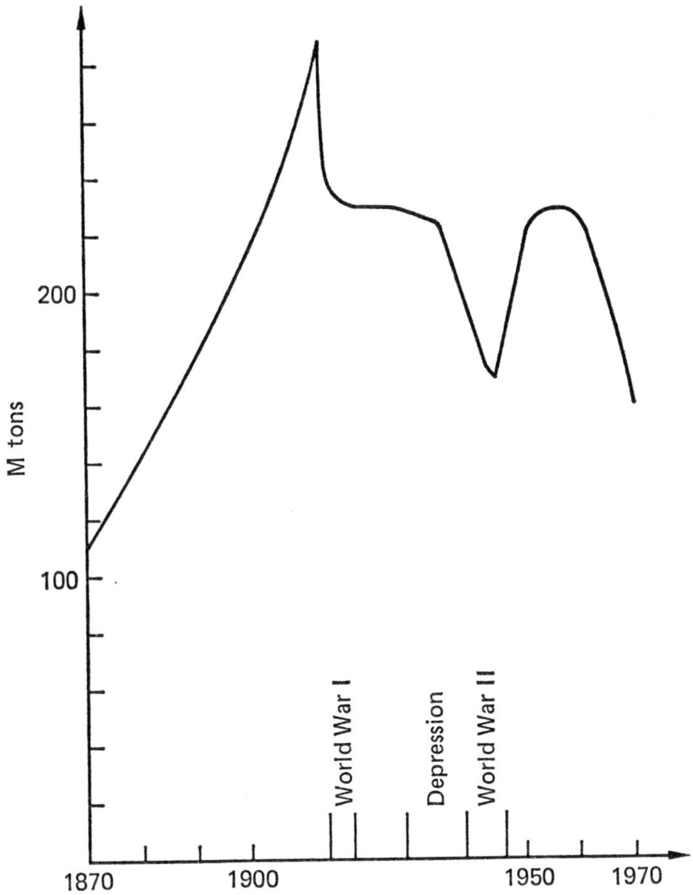

Fig. 4.10. *Production of coal in Britain, 1870–1970*

has been decreasing. This survey is taken from *The Revitalized Coal Industries* by H. E. Collins (*Colliery Guardian* publication, 1975).

Coal in the USA

The USA has enormous coal reserves: perhaps a total of 3×10^{12} tons, of which about one-twentieth are proved and economically recoverable. Even this large figure is not fully comprehensive, because it only includes coal seams which are sufficiently near the surface of land masses to be

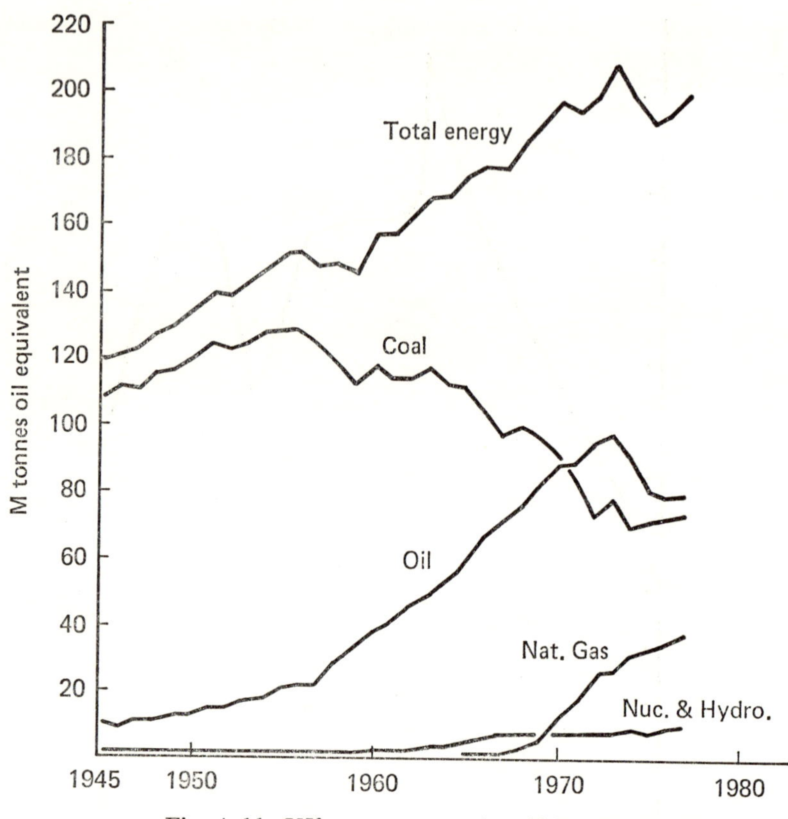

Fig. 4.11. *UK energy consumption, 1945–1977*

feasible for extraction by human miners. If one included all the coal at greater depths and far under the sea the figure might well be doubled or trebled. In the period from 1969 to 1974 there was a 30 per cent increase in the total energy consumption in the USA, the figure in 1974 being 2600 MTCE, which even with a population of 200 M corresponds to 13 TCE/person–year. Of this total, 45 per cent was oil, and nearly half of this oil is now being imported, 30 per cent came from the large reservoirs of natural gas, 20 per cent from coal, 3·9 per cent was hydro-power and 1·1 per cent was nuclear power. Even if the growth in total energy consumption in the USA could be stopped, the increasing competition for the earth's limited oil resources would make an increase in

United States coal production essential. The total reserves of coal suitable for strip mining are 5×10^{10} tons, that is, only one-sixtieth of the total reserves. Nevertheless, because of the much greater convenience of strip mining, some 50 per cent of the production in 1975 was strip mined, and there are plans to increase this proportion. The coal production had risen in the ten years before 1974, when it reached 600 M tons and the plans are to increase it to 1100 M tons by 1985.

Canada

The Canadian position is somewhat similar to that of the USA, with a steady increase in the total primary energy consumption to a figure of 250 M tons/year in 1974. Coal was only 11 per cent of this, oil 47 per cent, and natural gas 31 per cent. The coal consumption has remained substantially constant at 24 M tons/year, but the annual coal production increased from 9 M tons/year in 1969 to 22 M tons/year in 1973, and is by now at 25 M tons/year. The eastern coal fields have been worked for over a century, but the western coal fields are now steadily taking over. The total resources of coal are estimated at $0{\cdot}12 \times 10^{12}$ tons, 80 per cent of this being in the western coal fields of Alberta and British Columbia. In Saskatchewan there are sources of lignite suitable for strip mining and use in power stations.

Australia

The total coal reserves of Australia are $0{\cdot}2 \times 10^{12}$ tons, and the amount mechanically recoverable, by present methods, is 3×10^{10} tons. There is expected to be a considerable expansion in the use of coal for electricity generation.

South Africa

South Africa has traditionally mined its coal by the very wasteful 'room and pillar' method, which probably only recovers some 20 per cent of the coal seam. The coal production has increased steadily over the last thirty years, reaching 62 M tons/year in 1973. The total reserves of coal are estimated at 8×10^{10} tons, and the proved recoverable at 2×10^{10} tons. South Africa has been operating an oil-from-coal plant, the SASOL plant, for many years and, because of the high prices of imported crude oil, it is being planned to increase the size of this plant by 40 per cent from its present annual capacity of absorbing 4 MTC/year. The possibility of building a second plant is being considered.

India

Both in India and Africa the mining of coal has been increasing steadily. The total resources of coal in India are estimated at $0{\cdot}1 \times 10^{12}$ tons of which 2×10^9 tons is lignite. The coal has traditionally been obtained from deep mines, but large strip mines are now being developed for bituminous coals. The annual production of coal has increased from 76 M tons/year in 1969 to 81 M tons/year in 1973. It is planned to increase the production and use of coal in the country to 140 M tons/year in the early 1980s.

Japan

Primary energy consumption in Japan rose steadily after the Second World War to a figure of 360 MTCE/year, of which some 80 per cent was imported oil and coal (1974). One-third of this total annual primary energy consumption is used for generating electricity, but installed generating capacity is 50 000 MW. Seventy-two per cent of this is oil burning, 4 per cent coal burning, 3 per cent nuclear, and the rest hydro-electric. The total hydro-electric potential is estimated at 30 000 MW, and this will be completely harnessed by 1980. Japan is pushing on fast with the installation of nuclear power stations. Japan's total reserves of coal have been estimated at 2×10^{10} tons, but recoverable reserves at only one-twentieth of this figure for economic reasons. The annual output of coal from their own mines rose in the 1960s to a figure of 55 M tons/year, but was run down thereafter until in 1973 it was only 22 M tons/year, with some 50 M tons/year imported.

China

The total reserves of coal in China have been estimated at $0{\cdot}3 \times 10^{12}$ tons, and about one-quarter of this is regarded as proved reserves. However, the total coal resources, if we could find methods of mining them, are probably as high as $1{\cdot}0 \times 10^{12}$ tons. In 1958 the production of coal in China was 270 M tons/year and it is believed to be running now at 400 M tons/year. The total energy consumption of China is now about 500 MTCE/year so that 80 per cent of its energy comes from coal. China is already the third largest producer of coal in the world and it is expected that production will increase steadily.

Britain

The total consumption of primary energy in the UK has already been shown in Fig. 4.3 (with a population of 55 M). The total consumption in 1973 was just over 6 TCE/person–year. The total resources of bituminous coal and anthracite in Britain probably exceed 2×10^{10} tons, and the recoverable reserves are estimated at 6×10^9 tons. This does not include coal under the North Sea, which has been found in the recent drilling for oil, and which is probably equal to the whole of the resources under Britain. Coal was being run down in the period 1950–1973, but plans are now available to increase it by conventional mining methods in new modern mines to 150 M tons/year.

Germany

In 1974 the total primary energy consumption in West Germany was about 350 MTCE/year. Coal's share in this has been reduced in the previous five years from 40 per cent to 25 per cent and oil had risen to 55 per cent, most of this being imported; also, much natural gas was imported from the Netherlands. Electricity is generated mainly by thermal power stations fuelled by lignite and hard coal. Four per cent of the electricity output is nuclear, but it is intended to expand this rapidly. The total reserves of hard coal are 7×10^{10} tons, of which 3×10^{10} tons are regarded as recoverable. The total reserves of lignite are estimated at 6×10^{10} tons, of which about 1×10^{10} tons are regarded as recoverable. Annual production of hard coal reached a peak of 150 M tons/year in the 1950s, but it has now been reduced to less than 100 M tons/year. The annual output of lignite has risen steadily to a figure of 120 M tons/year; two-thirds of this goes straight to power stations adjacent to the mines. It is produced by highly mechanized strip mines. Lignite accounts for 33 per cent of the total electricity output while hard coal accounts for 30 per cent.

USSR

The total resources of all forms of coal in the USSR are estimated to be 6×10^{12} tons. The economically recoverable reserves of hard coal are given as 165×10^9 tons and lignite and brown coal as 107×10^9 tons. Hydro-electrical potential is estimated at 400 000 MW, of which at least half could be harnessed to hydro-electric generating capacity which in 1974 was 60 000 MW. The total annual production of primary energy in 1974 was $1{\cdot}5 \times 10^9$ TCE/year, of which $1{\cdot}2 \times 10^9$ TCE/year was consumed in Russia itself. The proportion of coal in the total energy supply declined from 50 per cent in 1963 to 30 per cent in 1973, but the actual annual consumption was increasing throughout this time. Estimates are put forward of production as high as $1{\cdot}0$–$1{\cdot}2 \times 10^9$ tons/year from highly mechanized strip mines and deep mines before the end of the century.

Poland

The total primary energy consumption in Poland was about 150 MTCE in 1974 and 80 per cent of this was provided by indigenous coal and lignite. The total resources of bituminous coal are estimated at 10×10^{10}

TABLE 4.2 WORLD COAL RESOURCES AND PRODUCTION
(from *World Coal*, November 1976, Vol. 2, No. 11, p. 37)

Country	*Economically recoverable, hard coal (×1 M tonnes)*	*Economically recoverable, lower ranks (×1 M tonnes)*	*Measured reserves, all ranks (×1 M tonnes)*	*Total resources, all ranks (×1 M tonnes)*	*Per cent of world resources*	*1975 hard coal production (×1 M tonnes)*	*Per cent of world production*
European and Asian USSR	165 800*	107 400*	73 204	5 713 681	53·0	536·0	22·4
Total North and Central America	128 320	59 680	377 912	3 045 280	28·2	593·2	24·8
Total Asia	114 668	1477	240 377	1 118 823	10·4	635·3	26·6
Total Europe	41 292	37 676	300 918	638 206	5·9	472·1	19·8
Total Oceania	13 805	10 709	74 694	199 641	1·9	68·6	2·9
Total Africa	12 335	315	26 607	52 029	0·5	74·9	3·1
Total South America	60	1814	3762	14 097	0·1	8·6	0·4
Total World	472 280	219 071	1 297 474	10 781 757	100·0	2388·7	100·0

Primary source: World Energy Conference Survey of Energy Resources, 1974.
Secondary source: World Power Conference Survey of Energy Resources, 1968.
Production data: International Coal Trade, US Department of the Interior, 4 April 1976. Vol. 45.
Hard coal includes anthracite and bituminous coals.
Lower ranks include sub-bituminous, lignite and brown coals.
Total resources includes all measured, indicated, and inferred resources.
* *World Coal* estimate.

tons, and half of this is regarded as economically recoverable. A quarter of this is within the reach of existing collieries and those under construction. Poland is now the fourth largest producter of bituminous coal in the world, with a figure of 160 M tons/year. By 1985 this is estimated to reach 215 M tons/year and by 2000, 300 M tons/year. Poland exports some 40 M tons/year of hard coal. The annual production of lignite has risen to 40 M tons/year and this is expected to increase still further. In the upper Silesian coal field mining takes place beneath heavily built-up surface areas and the wastes are completely stowed, together with sand, hydraulically conveyed from the surface, to eliminate surface subsidence. Seams as thick as 10 m are being worked by extracting 3 m at a time from the base of the seam and filling in by hydraulic stowing so that the next slice can be worked with the consolidated sand as a floor.

World coal resources

Table 4.2 shows the world coal resources and production by continents. The first two columns are the economically recoverable hard and soft coals, the third column the measured reserves of all ranks and the fourth column the total resources of all ranks. The global figure at the bottom of this column of $10{\cdot}8 \times 10^{12}$ tons is still only the resources estimated under the land at depths accessible to human miners. If we can find ways of extracting coal without men going underground, so that we can go far out under the sea and much deeper, this figure would be multiplied many-fold.

4.3.2 *Oil*

Oil is found in small spaces in porous rock and is obtained normally by drilling through a dome of impermeable rock under which it is trapped, sometimes with gas above and water below. The depth of wells has been steadily increasing and now we are also drilling beneath deeper and deeper seas. If the oil is forced to the surface by the pressure or can be pumped out this is called *primary recovery* and in the USA this recovers some 25 per cent of the oil in the reservoir. Secondary recovery means pumping water below the oil, or gas above it, to increase the pressure and drive it towards the well. Tertiary recovery can be achieved in some cases by lowering the viscosity of the oil, either by injecting steam to heat it or by introducing chemicals. The total recovery of oil in the USA is now about 32 per cent and it is hoped to increase it eventually to 40 per cent or more. The maximum annual rate of production from the world's proven reserves is not more than one-fifteenth of the estimated quantity in the reserves. This means that sooner or later we arrive at a point where

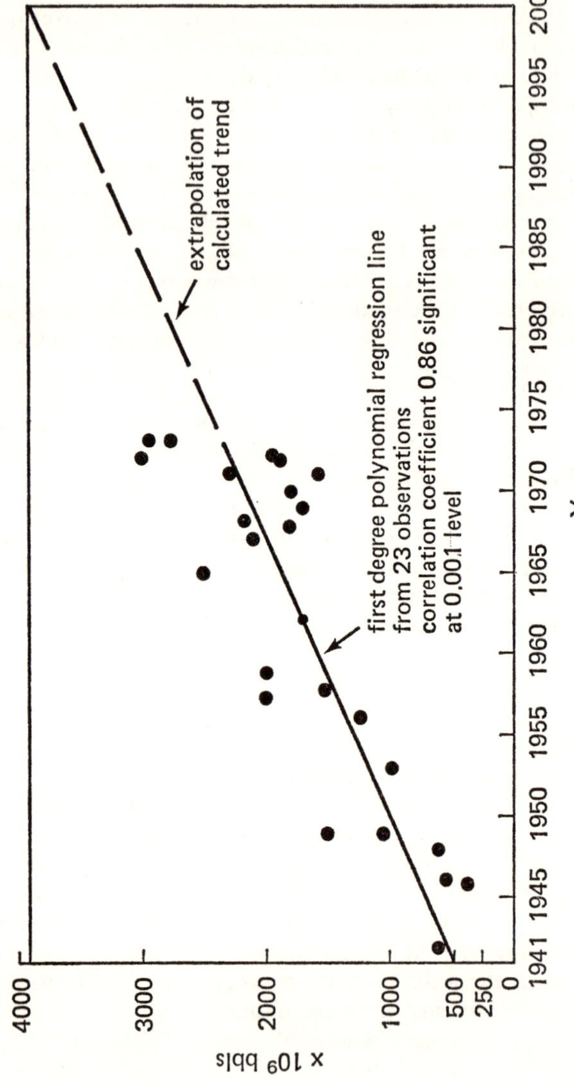

Fig. 4.12. *Calculated trend of estimates of world's ultimate oil resources (based on twenty-three estimates between 1942 and 1975). (From P. R. Odell and L. Valenilla,* The Pressures of Oil, *1978 (Harper and Row, London))*

the world demand is greater than one-fifteenth of the proven reserves, and at this point demand can no longer be satisfied.

Naturally a great deal has been written about the world oil reserves and resources and Fig. 4.12 shows the estimates of ultimate oil resources and how they have increased between 1942 and 1975. Taking five barrels as = 1 TCE, the figure of 2000×10^9 bbl corresponds to $0{\cdot}4 \times 10^{12}$ TCE. The Conservation Commission of the World Energy Conference obtained, in 1977, twenty-seven estimates of the total world oil resources, eighteen of these being from oil companies, and these averaged 300×10^9 tons of oil, which must be compared with 45×10^9 tons of oil already

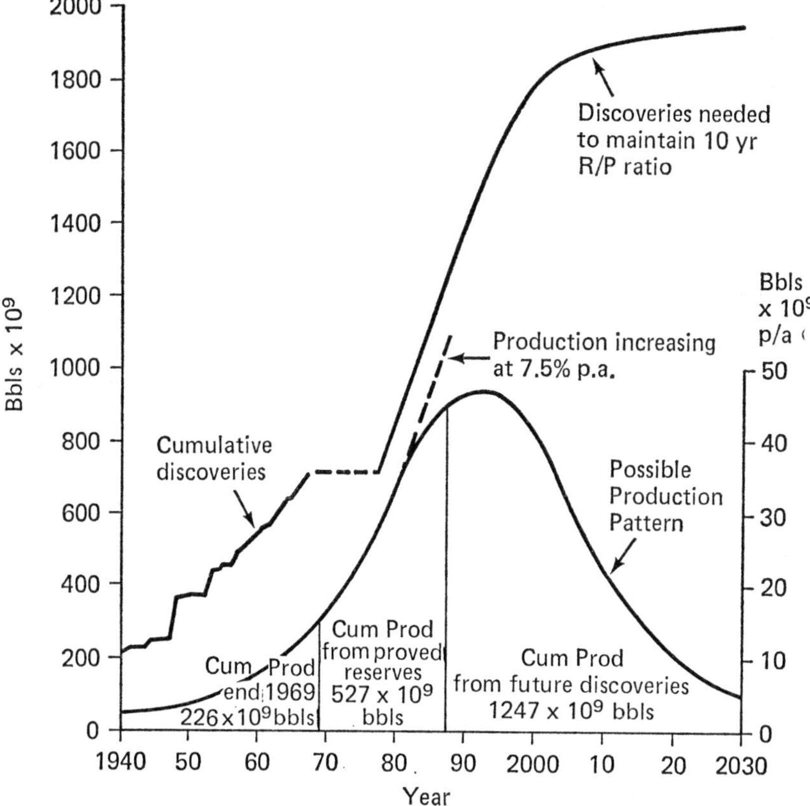

Fig. 4.13. *World oil reserves and production with 2000×10^9 bbl oil resource base and a 7·5 per cent/year rising demand curve. (From P. R. Odell and L. Valenilla,* The Pressures of Oil, *1978 (Harper and Row, London))*

produced by 1977. This figure corresponds to total oil resources of $0{\cdot}45 \times 10^{12}$ TCE, viz. one-fourteenth of the total coal resources.

Figure 4.13 shows the relation between cumulative production and cumulative discoveries, assuming a total world resources of $0{\cdot}4 \times 10^{12}$ TCE. This comparison indicates that world use will have to go down about the year 1995, and this corresponds to the view of most of the oil companies at the present time. Odell*, however, takes the much more optimistic view that one can extrapolate linearly the annual increase in estimates of total resources up to the total figure of $0{\cdot}8 \times 10^{12}$ TCE, and he assumes a substantially reduced rate of exponential growth from the previous figure of 7–8 per cent/year to a figure of 3 per cent due entirely to reduced growth in oil use in the developed countries as a result of the increased cost of oil. This argument appears to take no account whatever of the needs of the less developed countries, but as a result of his very optimistic figure for the total resources, and the slow growth rate curve, he arrives at a peak annual production of oil of 16×10^{9} TCE in the year 2025.

At the end of 1975 the proven reserves were $0{\cdot}11 \times 10^{12}$ TCE and some 80 per cent of this was in the member nations of the Organization of Petroleum Exporting Countries (OPEC), which is mainly the Middle East countries, but also includes Nigeria and Venezuela. In 1974 the proved recoverable reserves in the USA were 9×10^{9} TCE, but the indicated reserves were probably ten times as great and the total resources considerably larger again; in spite of this the USA is now having to import about half of its oil requirements. China has been developing her oil and natural gas resources over the last twenty years and is producing 75 MTCE/year. It could well become one of the major oil producers of the world. The USSR is also very well endowed with oil, with a current annual production of about 700 MTCE and total reserves as high as $0{\cdot}15 \times 10^{12}$ TCE.

4.3.3 *Oil substitutes*

The tar sands of Alberta constitute an enormous reserve of oil which will cost at least twice as much to produce as that which can be recovered in liquid form by pumping. It has to be extracted either by mining the sand and then heating it with steam, or by heating *in situ*; it also has the disadvantage of a very high carbon to hydrogen ratio, so that at least half of it remains from distillation processes as a petrocoke with a very high sulphur content. Odell estimates these reserves at $0{\cdot}4 \times 10^{12}$ TCE, but

* P. R. Odell and L. Vallenilla, *The Pressure of Oil*, 1978 (Harper and Row, London).

other estimates of that which will be recoverable at a reasonable price, are very much lower.

The geological survey of Canada has estimated the country's potential oil resources from tar sands as 70×10^9 TCE*.

There are oil shales in enormous quantities in very many parts of the world, but these contain at most a few per cent of oil and the only feasible way of extracting the oil at the moment is to mine them by conventional rock mining methods and then distil them in a retort. Up to the end of the Second World War shale was being regularly distilled in this way in the Edinburgh area. This will always remain a very expensive process.

Research has been carried out on making oil from coal since just after the First World War, and the main source of petroleum for the Germans in the Second World War was brown coal. In Britain the Billingham plant was put up between the wars to make oil by the hydrogenation of hard coal, but this was soon found to be uneconomic and they later made it by the hydrogenation of creosote produced as a by-product of coke manufacture. Hydrogenation is carried out with a catalyst at temperatures of the order of 450 °C and at very high pressures. The hydrogen for the process is produced by making water gas from coal and extracting the carbon monoxide. The other main process is the Fischer-Tropsch, which has been used in the South African SASOL plant. The coal is gasified with steam and oxygen and, after a further reaction with steam, produces a gas with about twice as much hydrogen as carbon monoxide. This synthesis gas is reacted with catalysts at a moderate pressure. This process, however, has a low overall thermal efficiency (about 30 per cent) whereas hydrogenation processes can achieve up to 70 per cent in terms of the calorific value of the oil produced compared with that of the coal input. Thus, the main extra costs of oil from coal, over and above the cost of the coal, equivalent thermally to the oil produced, are the coal which is lost in the conversion process and the capital and running costs of the plant. However, if mining methods can be developed to bring coal to the surface without men going underground this makes a very large additional amount of oil available for premium purposes after the pumpable oil has been exhausted.

4.3.4 *Natural gas*

Natural gas is frequently trapped over oil and obtained at the same time, although some wells give only natural gas, such as those in the southern part of the North Sea between Britain and Holland. About 80 per cent

* H. E. Collins, *The Revitalized Coal Industry*, 1975 (*Colliery Guardian* publication), p. 17.

of the natural gas deposit can be recovered from the borehole. In the Middle East the calorific value of the natural gas associated with the oil is between one-sixth and one-tenth that of the oil and in the past it has been flared to waste as it was not convenient to build special pipes to bring it away. Gas is far more costly to transport both by pipeline and by sea than is oil. More recently it is liquified and conveyed as LNG (liquified natural gas) in large tankers to industrially developed countries which can afford to pay for it.

TABLE 4.3. ESTIMATED RECOVERABLE RESERVES OF NATURAL GAS AND CRUDE OIL*†

	Natural gas	*Crude oil*	*Oil shale and tar sands*
Africa	8·4	23	4·2
Middle East	15·1	91	38
Asia	3·8	7·6	4·2
Europe	>6·3	>2·1	>4·2
USSR	25·2	14·7	6·3
USA	12·6	10·5	270
Canada	4·2	2·1	137
S. America	2·1	14·7	>2·1
Australia	>2·1	0·4	
World	82	166	460

* Taken from N. A. White, 'The cost of energy over the next decade', *Energy World*, November 1977, No. 42, p. 3.
† Units 10^9 TCE.

Table 4.3 gives estimates of the recoverable reserves of natural gas in various parts of the world although the estimate for Asia probably does not include an adequate figure for China. It is interesting that whereas the Middle East has some 80 per cent of the crude oil resources it has less than 20 per cent of the natural gas resources. Collins points out that the present proved reserves of the USA are sufficient for ten years supply, and he believes that when the offshore reserves are tapped there will be sufficient for forty years supply. Natural gas is an extremely cheap, clean*, and convenient fuel, and ideal for certain purposes, so it is quite certain that our descendants will consider that having flared it to waste in our haste to get oil out, or having used it up for purposes for which

* Where, as in southern France, it contains a lot of sulphur, this is easily extracted.

coal is a perfectly adequate fuel, is totally unfair to them. Possibilities other than piping it for direct use or for liquifaction are its conversion into methanol and other alcohols, which can then be conveyed conventionally as a liquid, and its use to produce electricity on site. A recent German proposal for off-shore oil wells with associated natural gas is to have floating gas-turbine stations sending electricity to the shore.

4.3.5 *Summary of position of fossil fuels*

Table 4.4 compares the estimates of the fossil fuels from the previous subsections, and Table 4.5 shows Britain's indigenous fuel resources.

TABLE 4.4 WORLD FOSSIL FUEL RESERVES AND RESOURCES

	Reserves (× 10^9 *TCE*)	*Total resources* (× 10^9 *TCE*)	*Resources recoverable by new methods* (× 10^9 *TCE*)
Coal	1300	10 800	21 000[1]
Oil	166	350	700[2]
Natural gas	82	200	500[3]
Shales and tar sands	460	900	10 000[4]
Uranium	17	43[6]	5 000[5]

[1] Assuming telechiric mining is developed to win seams down to 0·5 m thick, at depths up to 3000 m, and undersea distances up to 100 km from the shore.
[2] Assuming means are found of obtaining 90 per cent of all deposits.
[3] Assuming all natural gas is recovered and used from the most remote deposits.
[4] Assuming practical means are found for extracting all oils of significant concentration.
[5] Assuming successful breeder reactors and the extraction of uranium from granite.
[6] These figures are taken from 'World uranium resources', *Energy World*, June 1978, p. 2; assuming thermal reactors for which 1 ton uranium ~ 10 000 TCE, total resources at costs up to $130/kg uranium.

These fossil fuel reserves of coal, liquid petroleum, and natural gas took millions of years to lay down in the natural processes of plant life on the earth. Only a minute fraction of the sunshine arriving on the earth in these millions of years was trapped in plant growth and only a minute fraction of this plant growth was stored by nature. Nevertheless all the energy of these fossil fuels originally came from the sun.

It is clear that our present civilization will run out of these available liquid and gaseous hydrocarbons in less than a century overall. We have found ways of using them and propose to use them as fast as it suits us. In the case of coal it may last a few hundred years because the quantities are very much greater and particularly because the process of winning coal

TABLE 4.5 BRITAIN'S INDIGENOUS FUEL RESOURCES

	Years	*MTCE/year*	*Total MTCE*
Coal (conventional mining)	50	120	6 000
Coal (telechiric)	200	250	50 000
Colliery spoil (TCE)	50	15	750
Nuclear (fission)	30	50	1 500
Waste heat	200	50	10 000
Incineration	50	25	1 250
Solar	Permanent	100	
North Sea oil	25	150	3 750
North Sea gas	25	50	1 250
Nuclear (fusion)			
Tidal	Permanent	1	
Hydro	Permanent	5	

with human miners is very much less congenial and convenient. Such a study, combined with the observation that it is essential that the less developed countries, some two-thirds of the world population, should have all the benefits of the fossil fuels, leads one to three definite conclusions from the point of view of the long term interests of humanity.

(1) The fossil fuels, liquid petroleum and natural gas, are uniquely convenient for certain purposes and are in very much shorter supply than coal and they must, therefore, be regarded as *premium fuels* which must be used only for those purposes for which they are uniquely suited. For example, there is no known chemical substance which could compare with liquid hydrocarbons as a *fuel for transport* by air, road, or sea, and which can be obtained naturally in large quantities relatively cheaply*. The reasons for this are: (i) it has a very high calorific value per unit weight and per unit volume, (ii) it reacts with fourteen times its own weight of air, which it takes up from the surroundings as it goes along, and (iii) it is a liquid which can be carried in light tanks at ambient pressure and conveyed through small tubes.

It follows from this conclusion that it is a criminal offence, from the point of view of our descendants, to burn oil for purposes for which coal is a perfectly good fuel, such as power stations, domestic heating, or industrial firing. It is also, of course, a criminal offence

* The nearest is alcohol which is a convenient liquid with about half the calorific value.

to waste this precious *premium fuel* by using unnecessarily large engines or by burning natural gas off to waste.

(2) If the total energy consumption of the world is to be supplied mainly by coal for the next 100 years and the total world consumption of coal cannot more than double because of the carbon dioxide limitation discussed in section 4.2.1, then the world average per capita consumption must remain at about the present figure as the world population doubles. This can only be achieved *if the developed countries bring their average per capita figure down to the present world average of 1·8 TCE/person–year.*

(3) We need to double or treble the world production of coal in the next thirty years to provide all humanity (8000 M people) with 1·8 TCE and to supply all the uses where oil and natural gas are used at present and where coal can be readily substituted.

4.4. TELECHIRIC MINING OF COAL AND OTHER SOLID MINERALS

The fundamental reason why coal is more expensive to bring to the surface than oil is that men have to go underground to bring it up, and this requires safety, daily transport, ventilation, head room and all the other necessary conditions for human working. For some thirty years the nuclear industry has been working on remote handling systems because men could not come close to highly radioactive objects which had to be manipulated in connection with the processing of nuclear fuels and the refuelling of reactors. It is now possible to apply these remote systems to mining.

Telechirics means 'hands at a distance', that is putting the skill of a craftsman at the other end of a wire so that he can do the tasks of assembling, operating, maintaining and repairing machines from a safe, remote place. Automation has already gone a long way in regard to coal mining, but automated machines are liable to frequent breakdowns, as are the non-automated machines used for cutting and handling coal. As a result, in spite of three shift working and greatly improved conditions, continuous coal cutting machinery is only working 35 per cent of the time. When it breaks down a man has to crawl laboriously along, find the point of failure, and put it right.

It is not possible to ventilate a mine with high speed coal cutting and handling machinery to the point where a miner can work in it for a life time without getting some pneumonoconiosis. There are still accidents in mines from time to time, in spite of the very careful precautions which are taken nowadays. Some jobs in the mines, such as replacing broken cutters on a rotary drum shearer, have to be carried out in extremely uncomfortable positions. The daily journey of the miner to the coal face grows longer as the working faces get farther away from the shaft, and mechanization and automation has reduced the percentage recovery of coal from the 80 per cent which was at first obtained from long-wall mining, to some 40 per cent. Nowadays coal mining has to have shafts for transport and ventilation within a few miles of the working face so that a coal mining area is dotted with pit shafts every few miles, and it is not possible to obtain coal more than a few miles out from the sea coast. Finally, one must mention mining subsidence, which causes a great deal of damage to land and makes it impossible to mine coal under a town. As has already been mentioned, this can be overcome by pumping in sand and re-packing the spoil, but this a further expensive process.

The previous section showed clearly that there is twenty times as much

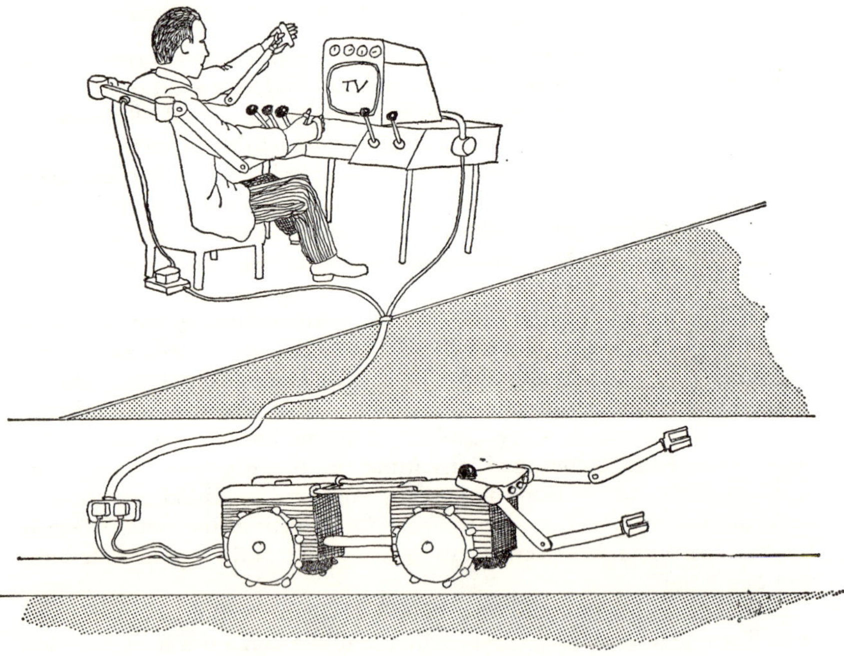

Fig. 4.14. *Diagram of telechiric mining*

coal under the ground as oil and natural gas put together, and that there will be very much more coal if we can overcome the limitations of mining with the human miner at the coal face. Figure 4.14 is a diagram of a telechiric mining operation. A body suitable for running about in a roadway no thicker than the coal seam (which might be down to 0·5 m) carries a pair of shoulders with arms and hands and some kind of three-dimensional view system, for example, two television cameras with eyes and a lighting system. The hands have force and touch feedback and the visual system is so arranged that the man on the surface can see the mechanical hands and feel what they are doing as though they were his own hands with gloves. He can handle any tools, and the hands can be ten or even 1000 times as strong as his own, so that he can pick up 1 ton as though it was 1 kg. For working in thick seams the arms can also be scaled up in length. As a first approximation one can imagine a number of these machines operating modern existing machines for cutting and transporting coal; ultimately, of course, these latter machines would be re-designed to integrate them in the working system with the telechirs.

The telechir, like the other machines, would be electrically operated so that it would have to carry power cables as well as the communication cables. It would, however, be possible for it to have a battery system so that it could move a considerable distance from one power point to another and plug itself in. It would also be possible to have a communication system with a modulated laser beam which it could pick up without being in mechanical connection. Another possibility for moving about would be for the operator on the surface to lock the telechir into a robot mode in which it unplugged itself and found its own way to the next power point, where communication would be re-established.

This simple visual description is only to show that there are no fundamental physical or engineering problems to which solutions are not already available. Much work has already been done on tactile feedback, proportional control of the seven movements required for a hand and arm with one grip from the human operator's hand, and three-dimensional picture transmission by stereoscopic television or by laser holography. (See Plate III.)

It would not be necessary to ventilate the mine at all so it could operate in a fuel rich atmosphere with no danger of explosion and with the fuel rich gas tapped off from below a gas lock at the surface containing inert combustion gases. A single drift would be sufficient to operate a whole coal field, even one extending fifty miles out to sea. This drift would have to carry: (i) conveying systems for coal (upwards) and fill material (downwards), and (ii) a machine-conveying rail track or roadway.

The telechirs would be operated by three shifts of men on the surface and would come to the surface once every three months for an overhaul. All the erection, operation, fault finding and repair on the coal cutting, road cutting, support, or conveying machines would be done by the telechirs *in situ*, as would the whole cutting and installation of the mine. The economics of coal winning would be enormously improved because the capital cost of the telechirs would be a small fraction of the capital cost of the double shaft and roadway system required for ventilation, and the multiplicity of shafts for man transport, all of which could be replaced by a single drift. It should also be pointed out that telechiric mining would be used for mining of all solid minerals, whereas underground gasification and combustion are only feasible for coal.

4.5. ENERGY GENERATION BY NUCLEAR FISSION AND FUSION

4.5.1 *The development of nuclear fission power*

The first fission of the uranium nucleus by means of a neutron was described and published just before the Second World War; since the fission process released more than two neutrons it was immediately clear that an energy releasing chain reaction was possible. As a result of the war both sides spent enormous sums of money working on the development of the nuclear fission bomb, and the Americans, with their allies, were successful. At the end of the war European countries had very well developed fuel economy drives, and the need for an expansion of energy resources was felt very keenly. A large number of brilliant physicists and engineers had worked on the fission bomb and the idea that uranium fission could provide unlimited cheap electricity in two or three decades was sold to the public and the political decision makers. The production of materials for fission bombs was the hidden argument. As far as Britain is concerned this has meant that enormous establishments spending tens of millions of pounds a year have been working for the last thirty years in the hope of carrying out this dream. Almost every other line of research in applied science has been relatively starved of funds, and an establishment has been set up with a vested interest in carrying on this work. A great deal of money has been spent on propaganda and it has been very difficult in this regard for those looking at this unbalanced development from the outside to obtain the facts, both because of secrecy, and because for them it is a spare time, non-funded activity. Although the purpose of this book is to try to help the engineer to make up his own mind about the ethical or moral problems in engineering, without bias, it is necessary to express a definite bias against nuclear energy because so much time and money has been spent with a bias in its favour*.

In 1957 Britain launched the Magnox programme of relatively low temperature gas cooled reactors. Nine stations were built with a total energy capacity of 3800 MW and these saved 12 M tons of coal in 1974. However, in 1965 a programme to build five Advanced Gas Reactors (AGRs) was started but none was working in 1975. Looking back over the last thirty years it is hard to avoid the impression that in terms of the present state of engineering and the present benefit to mankind this enormous expenditure on research and development has been almost entirely wasted and that it now continues primarily by its own momentum. At the present time the primary arguments for the continued expansion of nuclear power come from two groups of people, firstly, from those that feel that the developed countries need an ever continuing exponential growth in their use of electricity and, secondly, from those countries that feel that nuclear power stations will give them the opportunity to manufacture their own bombs. In Britain the annual vote for nuclear research and development is of the same order as all other government supported civilian research and development put together and is not subject to Parliamentary scrutiny because some of the plutonium is required for military purposes.

The first AGR, Dungeness 'B', was ordered in 1965 and has given considerable trouble, so that it will not now be completed until at least 1980. Four more AGR stations have been constructed, two of them being commissioned in 1976, four years behind schedule, and the other two when completed will be more than three years late. All five stations have cost very much more than originally estimated and have had major operational troubles. One of those commissioned in 1976 suffered a major malfunction in July 1977 and has been shut down ever since. At the present time Britain has electric generating capacity of over 65 000 MW and is constructing a further 16 000 MW, while the highest peak demand is less than 60 000 MW. The peak demand has increased by less than 3000 MW in the last six years. Britain is prematurely shutting down smaller stations to justify the large new ones it has built. There is a new line of thought that the economies of scale claimed for the large 1000 MW stations is not correct and that the optimum size is probably 120–150 MW. Such small stations can readily be combined in total energy systems (pass out steam used directly for heating). Also, large stations are very clumsy in reacting to rapid changes of demand.

* For example the CEGB advertises free talks on nuclear energy in the Women's Institute journal.

One very significant fact is that a number of senior engineers and scientists who had had many years of experience in the nuclear industry, especially in America, have been leaving it and writing against it. David E. Lilienthal, former Chairman of the United States Atomic Energy Commission, wrote in 1972: 'Once a bright hope shared by all mankind, including myself, the rash proliferation of atomic plants has become one of the ugliest clouds overhanging America'.

E. Broda* describes thirty-five years of intimate connection with nuclear energy research which have led him to say: 'I began as an enthusiast of the peaceful applications of nuclear energy. However, gradually I developed a critical attitude. My scepticism is now, although I still do not consider myself an absolute opponent of nuclear energy, quite pronounced.' He ends with a plea for a crash programme for solar energy research.

In the *New Scientist*, 12 February 1976, there is a reference to three senior engineers who had left General Electric and who are quoted as saying: 'There is no way you can continue to build power plants and operate them without having an accident'.

G. L. Weil† who was a research associate of Dr Fermi, and took part in the achievement of the first nuclear chain reaction, resigned from the United States Atomic Energy Commission in 1952. In March 1974 he wrote an article entitled 'It ought not to have happened in the way that it did'. The following two paragraphs are extracted from this article.

> Comparisons of safety records with other industrial operations are very misleading, not only because, by contrast, the nuclear industry is very young, but more importantly, because the effects from radiation exposures, whether instantaneous or cumulative, do not necessarily show up immediately as countable corpses or visible injuries. A commercial plant of 1000 MW will contain the radioactivity equivalent to the fall-out of 1000 Hiroshima type bombs. Breeder concepts will be even more hazardous because, in addition, large amounts of plutonium will be used as the nuclear fuel. In this context, nuclear fission energy is not the 'cleanest and safest', but rather the 'dirtiest and most hazardous' form of energy known to man.
>
> A USAEC study performed eight years ago, but with conclusions released only recently, of a hypothetical major accident at a then 'modern' US 800 MW nuclear plant involving the release from the containment to the atmosphere of an arbitrarily assumed 15 per cent of fission products could cause 45 000 fatalities,

* E. Broda, 'Long term problems of nuclear energy', Paper, Institute of Physical Chemistry, University of Vienna, Austria, December 1976, p. 429.

† G. L. Weil, *It ought not to have happened in the way that it did*, reprint from *Towards Survival*, March 1974.

74 000 injuries, contamination of 50 000 square miles and $17 billion damage. There was no estimate of the genetic damage which would be inherited by future generations.

Drs J. W. Gofman and A. R. Tamplin*, research associates at the Lawrence Radiation Laboratory at Livermore, California, have unrivalled knowledge of the biological effects of radioactivity and they argue that if the allowable exposure set by the Federal Radiation Council of the USA was reached it would create, in the USA, an additional 32 000 cases of cancer and leukaemia per year. They also argue that the Atomic Energy Commission has not funded safety research at an appropriate level, and that present reactors, and those under construction, are far more experimental than we might have imagined.

In a Declaration on the 30th Anniversary of Hiroshima, J. T. Edsall, H. W. Kendall, G. B. Kistiakowsky, H. C. Urey and J. D. Watson, all American physicists, chemists, or biologists of the first rank, wrote:

> It was no mistake, following Hiroshima, to try to make use of nuclear energy for peaceful purposes. But it was a serious error in judgement in the following decades to devote resources to nuclear development to the virtual exclusion of other alternatives. It has also been unfortunate that the efforts to commercialize nuclear energy allowed safety and national security problems to receive less than the required consideration. The nation, on the thirtieth anniversary of Hiroshima, must take note of these facts, diminish the large growth rate of the nuclear program, and take other appropriate steps to ensure adequate energy for the nation.

The recent escape of radioactivity due to misoperation of the nuclear power station near Harrisburg, USA, is an example of how much more widespread are the effects of a nuclear station accident than any other industrial accident.

Robert Jungk whose classic book on nuclear weapons, *Brighter Than a Thousand Suns*, has been read with dread all over the world, has recently written *Nuclear State*†. In this he concludes that the social difficulties raised by nuclear power are insurmountable. These arise from the dimension of the hazard (the number of people who might be killed by a nuclear accident) and the harm to future generations from accidents. There is a whole chapter about the intimidation of critics.

* J. W. Gofman and A. R. Tamplin, *Poisoned Power, the Case Against Nuclear Power Plants*, 1971 (Rodale Press, Emmaus, Pa.).

† Robert Jungk, *Nuclear State*, 1979, translated by Eric Mosbacher (John Calder, London). See also Michael Kenward, 'A thousand times gloomier', *New Scientist*, 5 April 1979, p. 7.

4.5.2 *Uranium fission*

When a low energy neutron hits a ^{235}U nucleus it causes the nucleus to break into two parts of roughly equal size and to release an average 2·5 further neutrons at very high energies. In order to have a stable chain reaction it is necessary that one of these neutrons is slowed down to low speed before it is stopped by entering another ^{235}U nucleus. This is done by placing the uranium in rods of quite small diameter supported by a moderating material which slows the neutrons down before they come to another rod. The chain reaction is controlled by adjusting rods of neutron-absorbing material in the reactor to keep the reaction exactly stable at the required level. By a fortunate accident a small proportion of the reactions are much slower which allows time for mechanical control, otherwise the reaction would accelerate to the point where the system blew up, as it does in the nuclear bomb. The heat is taken out with carbon dioxide in British reactors (Magnox and AGR) and is carried to a boiler. In these reactors the moderator is graphite and the uranium is enriched to 0·672 per cent of ^{235}U. The other reactors use ordinary water or heavy water as the moderator and the heat removal system. The water must be in liquid form in the reactor, but can be allowed to boil for power generation or it can be used as a heat exchange material to generate ordinary steam. As the fuel concentration of ^{235}U goes down the fuel element is removed by very elaborate and expensive handling arrangements, and at this stage it can be processed chemically to separate out the plutonium (Pu) which has been formed by the impact of high energy neutrons on the ^{238}U; this plutonium can then be used as a fissile material for further power generation or, of course, for making bombs. As the re-processing plant is, in general, far from the power station it involves the carriage of highly radioactive spent fuel by sea, rail, and road, inside heavy lead containers, and further very complicated handling at the re-processing plant.

Since the amount of relatively easily won ^{235}U is so limited and the process of concentrating ^{235}U from natural uranium, by purely physical methods, is very expensive because there is only about a 1 per cent density difference, the fast breeder reactor is being developed. The fast breeder reactor can use up to 100 times as much of the uranium and can also use thorium, since it can convert these into fissile nuclei. The fast breeder reactor has a central core of highly enriched fissile material cooled with liquid metal and surrounded with a blanket of fertile material (uranium or thorium) so that fast neutrons escaping from the central core can convert some of the blanket into fissile nuclei. It is possible to adjust this system so that it consumes more fissile material in the core than it makes in the blanket or the other way round. The blanket has to be reprocessed chemically to extract the fissile material, again with the associated problems of transporting and handling highly radioactive material. About 3 tons of highly radioactive fission products have to be disposed of from a year's operation of a 1000 MW station.

4.5.3 *Resources of uranium*

Uranium-235 is the only convenient fissile material available in nature and is present to the extent of 0·7 per cent in natural uranium, whereas ^{238}U is the principal constituent and absorbs neutrons to make heavier elements such as plutonium, which is itself fissile.

If one could cause fission of the whole of a ton of uranium it would produce energy corresponding to about 2 MTCE. The maximum amount of energy one can get out of a ton of highly enriched uranium corresponds to about 1 MTCE, while the maximum amount one can get of slightly enriched uranium corresponds to about 10 000 TCE. Recent estimates of the known world reserves are that we have 1·65 M tons of natural uranium which would be sufficient for twenty years forward requirements at present growth rates. Beyond 1985 there will be an increasing need to find, conventionally, new sources if we continue to expand nuclear fission energy. The only countries of the world which were known to have more than a thousand tons resources in 1977 were Canada, France, Niger, South Africa, and the USA; the quantities in the USSR are not known. Thus if one equates each ton of this uranium to 10 000 tons of coal one obtains a total resource of uranium for energy production of $1{\cdot}7 \times 10^{10}$ TCE, something like one-sixtieth of that of the readily available solid fuels and one-tenth of that of the available crude oil of the world. (See Table 4.4.) If, however, one is prepared to spend very much more on obtaining uranium one can crush all the granite mountains of Norway and Sweden and other countries and extract the uranium—at enormous energy cost, and with severe damage to the environment.

4.5.4 *Nuclear energy problems*

(a) *Capital cost.* For a nuclear energy power station to be viable it has to be at least 500 MW and preferably 1000 MW and has to be on base load, that is, essentially, to be running for as much of the year as possible. These two considerations mean that to set up a national electricity system, if it is to have even one nuclear station, it has to contain several large power stations of which at least one-third must be non-nuclear to cope with the peak demands. The system has to be integrated with a national grid which involves further enormous capital costs. A recent paper* gives the following construction costs as at January 1977.

Coal: £290/kW, equivalent to £580 M for a 2 GW station.
Nuclear (SGHWR): £520/kW, equivalent to £1050 M for 2 GW
Nuclear (AGR): £470/kW, equivalent to, £940 M for 2 GW
Nuclear (AWR): £407/kW, equivalent to, £814 M for 2 GW

These are for the power stations alone.

The overall cost, of course, depends on the amortization rate for capital, and on the fuel costs, which are considerably higher for coal at the present time, although it is doubtful whether nuclear has been correctly charged for research and development costs and for fuel processing. An earlier assessment by D. Leslie† compared, in 1973, the capital cost of a nuclear power station in the region of £150–250/kW, with oil capital cost at £76/kW, but when the fuel costs were added in he obtained exactly equal overall costs. The fuel cost was made up of three equal factors, the cost of natural uranium, the cost of the enrichment of natural uranium to 2/3 per cent ^{235}U, and the cost of fabrication and processing. The cost of waste disposal is not, as yet, adequately included.

These capital costs for a nuclear power station are extremely high, and are continually rising, as more difficulties are encountered which are overcome by more expensive designs. Peter Chapman and John Price of the Open University have pointed out how much energy is required to build a nuclear power station, to mine, separate and process the uranium and to provide the other necessary materials. If one is considering the supply of energy to an undeveloped country one must take account of the need to build the power stations in that country, set up a complete electricity distribution system and provide all the back-up facilities, such as transport for fuels and spent fuel elements. Chapman has also shown that if the number of reactors in a country is to double every 4·3 years then less than half of the energy output of a station is available for other purposes, the remainder being required to install the additional power stations. One is forced to the conclusion that *nuclear energy is, at the very best, a 'rich man's toy'*, that is, a way of enabling the rich countries to continue their very extravagant use of energy; *it could not possibly provide enough energy for the poor countries*, nor is there any sign whatever that more oil will be made available in the poor countries as a result of the rich countries installing nuclear power stations.

Another fundamental problem with nuclear power stations is that

* 'Coal and nuclear power station costs', UK Department of Energy, 1977.
† D. Leslie, private communication.

nobody wants them near to their homes or factories, and this makes it very difficult to combine them with pass-out steam heating systems, so that they will always be condemned to an overall thermal efficiency of the order of 30 per cent.

All these arguments about the size and capital cost of nuclear power stations and the associated distribution systems apply equally to nuclear fusion, which will almost certainly have to be on a scale of 2000 MW and run as a base load power station.

(b) *Problems of radioactivity and human safety*

> Exposure to nuclear radiation can cause cancer, it can cause babies to be born mentally or physically defective, and it can cause increases in many serious illnesses like heart disease. I know of no one who denies these statements.
>
> Fortunately, there is a chance to prevent serious nuclear pollution; it has not yet occurred. The threat, however, lies in the country's growing commitment to nuclear power plants for electricity, and to nuclear weapons for defense.
>
> The problem with nuclear electricity is that as much long-lived radioactivity is produced inside one large nuclear power plant every year as there is in the explosion of about 1000 Hiroshima bombs.
>
> When we say 'long-lived' radioactivity, we mean long. Some kinds last for 100, 300 and even 240 000 years before decaying fully.
>
> Unprotected, above-ground nuclear power plants, loaded with radioactivity in their cores, would certainly be large liabilities if this country were ever under attack. They seem to make the country virtually indefensible*.

Although these problems have been well publicized they have also been vigorously rejected by the nuclear establishments to whom the figures are uniquely available. This is, of course, a very dangerous situation since the nuclear establishments can say '*we* have calculated that the risks are minute', but the people concerned about it do not have the figures to check these calculations. They publish calculations of risks of power stations accidents in peacetime but ignore the problems of transporting and handling radioactive materials, and of war and sabotage. There is a great deal of evidence that many serious accidents have not been reported until a long time afterwards.

The best known problem in Britain is the disposal of the radioactive waste. At present, after reprocessing to recover plutonium and the unconsumed uranium, the nuclear fission products and the transuranic elements are extracted by dissolving in nitric acid so that the highly

* Mike Gravel (US Senator from Alaska), from the Foreword to J. W. Gofman and A. R. Tamplin, *Poisoned Power, the Case Against Nuclear Power Plants*, 1971 (Rodale Press, Emmaus, Pa.).

radioactive elements are reduced to an acid solution of small volume which is stored at the Windscale processing plant as a liquid. One GW year of electricity produced 14·5 m^3 of highly active waste, and some 700 m^3 have been accumulated in this country. This waste is stored in tanks which require maintenance so that, at the present time, we are leaving a material for our descendants to look after for several hundred years, at their peril, which is of no use whatever to them. Some research has been done on turning it into glass and sealing it up in salt mines or other rock formations, but a satisfactory disposal method has not yet been found, neither has a method of eliminating the effects of the heat which will be produced for many years by the radioactive decay.

Another major problem which has not yet been solved is what to do with derelict nuclear power stations. They will have an operating life of some twenty-five years, and the cost of dismantling them would be very high indeed with no benefit to be obtained from it, thus *we are condemning our descendants to live in countries with untouchable mausolea.*

The United States Environment Protection Agency (EPA) has recently promulgated standards for emission of radioactivity which are 100 000 times stricter than the British equivalent. Professor Carol Morgan observed at the Windscale enquiry that the Magnox plant would never have been allowed to operate in the USA because it fails to meet discharge standards. The standards in Britain are based on research carried out by employees of the UK Atomic Energy Authority (UKAEA). They admit that there is a risk to people, due, for example, to inhaling alpha particle emitters, but they attempt to balance this against the necessity for more electricity, a truly Faustian bargain, as admitted by Alvin Weinberg, the American expert and advocate of nuclear power.

> We nuclear people have made a Faustian bargain with society. On the one hand we offer an inexhaustible supply of energy. But the price we demand of society for this magical energy source is both a vigilance and a longevity of our social institutions that we are quite unaccustomed to. . . . We have relatively little problem dealing with wastes if we can assume *always* that there will be intelligent people around to cope with eventualities we have not thought of.

When one considers that in fact we cannot sell all the electricity we make, at the present time, and that most of the electricity is used for unsuitable purposes, like heating water or making throw-away plastic containers, then it becomes clear that *the risk to human health, however small, should outweigh the need for more electricity.* John Davoll, in an article 'The case against nuclear fission power'* points out that, if

* John Davoll, 'The case against nuclear fission power', *Laser*, September 1974, p. 10.

UK plans are fulfilled, in twenty-five years time our power stations will generate each year as much radioactivity as the fall-out from 100 Hiroshima-type bombs, and this is in Britain alone. He says:

> Let us be a little more specific about what the situation will be in about 30 years' time, if current plans are fulfilled. There will be several thousand reactors around the world, many of them plutonium-fuelled fast breeder types. Plutonium sufficient to make several hundred nuclear bombs will be in transit at all times. Reactors and fuel processing plants will contain so much radioactive material that the dissemination of a tiny fraction of it would render large areas uninhabitable for centuries, and if the installations are sabotaged or destroyed in war, or their staffs killed or driven away, such dissemination is more than likely. Moreover, the longer the social system survives, the larger will be the amount of radioactive materials at risk as the undeveloped countries follow in the footsteps of the rich. Once nuclear fission has become a major contributor to energy supply, no war or large scale disorder will again be permissible.

Hottel and Howard* pointed out in 1971 that 'the amount of radio active emission occurring during reprocessing is about 100 times larger than that occurring at the power plant, and the long lived radionuclides ^{85}Kr and ^{3}H are released in their entirety'.

All experience shows that it is not the accidents which are expected which occur, but those which no one expected, which arise often because of human error, carelessness, or even human malignance. For example, it has been shown that the danger of a reactor core melt-down can be reduced to negligible levels in peacetime. If a country having nuclear power stations is at war with another country, the situation is quite different. It has been calculated that if all the nuclear power stations at present in France were to be blown up by conventional high explosive weapons, almost the whole of France would be covered with radioactive fall-out. Even in peacetime there are acts of sabotage, hi-jacking, and terrorism which make the question of transporting plutonium so critical that it is necessary to support the nuclear activity by the operation of the army or armed police. The Royal Commission on Environmental Pollution (Chairman, Sir Brian Flowers) reported in 1976; three of its conclusions are:

> Plutonium appears to offer unique potential for threat and blackmail against society because of its great radiotoxicity and its fissile properties.

> There should be no commitment to a large programme of nuclear fission power until it has been demonstrated beyond reasonable doubt that a method

* H. C. Hottel and J. B. Howard, *New Energy Technology*, 1971 (MIT Press, Cambridge, Ma, USA), p. 239.

exists to ensure the safe containment of long-lived, highly radioactive waste for the indefinite future.

The dangers of the creation of plutonium in large quantities in conditions of increasing world unrest are genuine and serious. We should not rely for energy supply on a process that produces such a hazardous substance as plutonium unless there is no reasonable alternative.

E. Broda has argued* that nuclear power plants and fuel reprocessing plants could not be kept under reliable control under war conditions so that lethal radioactive escape over large areas would occur even without the direct bombing of the reactors. For example, if the cooling and emergency cooling systems stopped 'billions of curies, equivalent in activity to thousands of tons of radium, could eat their way out of each power station affected'. A 1 GW plant produces three times a day as much radioactive fission material as the explosion of a 1 kg uranium or plutonium bomb.

4.5.5 *Nuclear fusion*

In theory nuclear fusion offers mankind the possibility of almost unlimited power since it is theoretically possible to make two heavy hydrogen nuclei (deuterium D) combine into one helium nucleus, with the conversion of a significant fraction of the total mass into energy. Heavy water (deuterium oxide) is present in all natural water at a concentration of about 200 ppm; the total fusion of the deuterium in one gallon of water would produce more heat than is obtained from chemical burning of 300 gallons of petroleum. The basic problem with nuclear fusion is that the particles have to be heated to an energy equivalent to a temperature of some hundreds of millions degrees kelvin, and it is, of course, immensely difficult to contain a gas at such a temperature owing to the fourth power law of thermal radiation. There is, however, one reaction which is 100 times faster than the D–D reaction, this is the one between deuterium and tritium (D–T reaction). Tritium is the still heavier isotope of hydrogen and does not occur in nature, it has to be made by the neutron bombardment of ^{6}Li which forms 4·7 per cent of natural lithium. The energy content of a ton of ^{6}Li is not very different from that of a ton of uranium. Natural lithium is at least four times as abundant as uranium but only one-thirteenth of it is ^{6}Li.

The immense interest in the fusion reactor is not so much because it gives us much more energy but because the total radioactivity is minute

* E. Broda, 'Destruction of nuclear power plants in crisis and war', 28th Pugwash Conference, September 1978, Varna, Bulgaria.

compared to that in a fission reactor. There are no risks of explosion or loss of coolant and there are no explosive materials for people, so disposed, to steal. Unfortunately, a self-sustaining fusion reaction is only just on the point of being achieved in the laboratory. If we compare this with the development of the fission reactor the first laboratory reaction was achieved in September 1942 and now, nearly forty years later, commercial fission reactors are still showing endless teething troubles.

The two methods which are being studied in the attempt to achieve conditions for fusion are: (i) magnetic containment using an enormous toroidal magnet to produce a fine ring of plasma in the centre of a large vacuum, (ii) to focus a spherical laser pulse on to a spherical pellet of fusile material. Both these methods have been worked on for more than ten years and it is by no means certain, as yet, that it will be possible to produce a sustained power output and feed it into an electric distribution system. What is quite certain is that it will require an enormous reactor and correspondingly enormous distribution system, so that the actual cost of the electricity will still be very high, because of capital costs.

4.5.6 *Conclusion*

One is forced to the conclusion that the possibility of obtaining cheap electricity from nuclear energy, which was believed just after the war, has proved a chimera, because, although the fuel is cheaper, the cost of the generating plant and associated equipment has proved to be very much higher than the corresponding costs for thermal power stations. One must, therefore, conclude that *electricity is the premium fuel par excellence* because not only does it require very high capital cost to make it, but also the overall thermal efficiency, whether made from chemical fuels or from nuclear fission, is less than 30 per cent by the time it gets to the user.

A good example of the change in ideas is given by a paper written by Dr Haber, then chief fuel officer of ICI, entitled, 'UK fuel and power'*. In 1955 our total energy consumption was about 230 MTCE, almost all of this coming from coal. By 1980 he predicted that it would be 360 MTCE made up of 240 M tons coal, 35 MTCE oil imports, 2 MTCE hydropower and 90 MTCE of atomic energy. In fact the situation is as given in Table 4.6.

4.6. RENEWABLE ENERGY RESOURCES

4.6.1 *Water power*

Water power obtained from rainfall in hills or mountains was one of the

* *J. Inst. Fuel*, February 1957, **29**, 65.

very first forms of energy to be available to man, with the water wheel. Water turbines have been developed with very high efficiencies for power generation. For more than thirty years water systems have also been used for energy storage, so that during periods of low energy requirement one set of turbines could drive, electrically, another set to pump water back into a high reservoir, which was then available for peak energy requirements. It has been estimated that the total potential water power

TABLE 4.6 ENERGY USAGE IN BRITAIN (MTCE)

	1973	*1976*	*1978*
Coal	131	120	120
Gas	43	58	68
Oil	162	132	132
Nuclear and hydro	12	15	19
Total energy uses	348	325	339
Non-energy oil	20	17	20
Indigenous production			
Coal and gas	185	193	197
North Sea oil		20	108

Taken from Energy Commission Paper No. 3, Department of Energy.
1973 was the year of peak energy consumption.

resources of all the mountains of the world is about 3×10^6 MW, which is approximately equal to the present world's industrial electricity supply. At the present time the world has about 250 000 MW installed hydro-power capacity, of which 45 000 is in the USA, where 75 per cent or more of the potential water power is already being used. The basic problem of water power arises from the enormous amount of water required. For example, a 1000 MW power station can operate with normal thermal efficiency on 360 tons of coal per hour but would require $4{\cdot}6 \times 10^5$ ton/km of water per hour at 80 per cent conversion efficiency. The capital cost of hydro installation depends enormously on how conveniently nature has placed a river (in a narrow gorge in high mountains so that a small dam can produce the required head is ideal), but in all cases it is much higher in capital cost than a thermal power station. There is, moreover, the problem of the upsetting of the normal flow of silt in the river which becomes deposited in the reservoir so that it is full within 50–100 years. There are also, in certain cases, very considerable problems arising from the flooding of valleys of natural beauty and human utility, and there may even be problems of earthquakes due to the enormous weight of impounded water.

Another form of available water power is tidal energy. The French have installed a very well engineered dam across the Rance Estuary which gives 240 MW for about one-third of the time, operating both during tidal inflow and tidal outflow. It has been estimated that the world total of basin sites is 64 000 MW, roughly 2 per cent of the world's total possibilities for hydro-power. The possibility of a tidal barrage in the Severn estuary was first considered in 1933, but the high capital cost has always made it uneconomic in periods of cheap fuel. It would have saved 10^6 tons of coal every year since then if it had been installed, and would probably have brought Britain out of the Depression some years before the Second World War. It is now being reconsidered as the Severn estuary is a very suitable site because the tide is focused into a narrow estuary and by means of suitable barrages it would be possible to obtain 2–5 MTCE of electricity every year. It is also possible to use it to provide pumped storage to give the power when it is wanted. It would take twenty years to build and it is very difficult to estimate what effect the barrage would have on the head of water available.

The third water energy resource that is being investigated is the use of waves. Waves represent a way of collecting some of the wind power across an enormous stretch of ocean. It is estimated that the energy in the waves ten miles off-shore from the north of Scotland is as high as 70 kW/m so that 400 km of wave collecting apparatus, with a reasonable efficiency, could save 25 MTCE/year, at present used for electricity generation. The conversion devices have to be placed many miles out to sea since the waves lose their energy as they come to the shore, and various devices, in the form of enormous articulated concrete rafts, are being studied. In some the water flow is rectified to drive water turbines continuously in one direction, in others air turbines are used. Some estimates of £800/kW have been put forward for the capital cost (in January 1977), but clearly there are enormous problems to be overcome in bringing the electricity, or electrolysed hydrogen, to the shore, in anchoring the system in all wave conditions, and in the general problems of life and servicing.

4.6.2 *Wind energy*

Windmills have also been traditionally used for some hundreds of years for pumping water and grinding corn. A large four-bladed windmill was highly developed in this country and in Holland and could give up to about 10 hp. Early in the last century it was proposed to pump dry the

Haarlemmermeer in Holland with 150 windmills. These, however, were replaced by a single 12 ft diameter atmospheric steam engine working eight beam pumps.

The principal problem of wind energy is that it is very variable and the power output depends on the cube of the wind velocity, U, according to the formula

$$P = C_p \tfrac{1}{2} \rho U^3 A$$

where ρ is the air density (1·25 kg/m^3) and C_p is the coefficient of conversion of the wind energy in the swept area, A, which is in the range 0·15 to 0·45. The average wind power depends, therefore, on the average of the cube of the velocity, so that it is very critical to use an area where the wind is above average, such as the top of a hill, and detailed contours of higher wind availability have been constructed for Britain and the world. For windmills rated at 11·25 m/s, most of Britain is above 2250 annual kWh per kW installed capacity, and the coastal areas are better than inland, particularly the west coast. An advantage of wind energy is that the mean wind speed is relatively constant from year to year, $\pm$10 per cent. Another advantage of wind in Britain is that it is a maximum in the winter when there are greatest needs for heat and power, whereas solar energy is a maximum in the summer. The wind speed varies with height, h, above ground level according to the formula

$$U \propto h^{\alpha}$$

where α lies between 0·1 and 0·3. It is necessary to have the windmill at a considerable height, and if it is a horizontal axis wind turbine there will be a cyclic variation as each blade comes to the lowest point.

Windmills are basically of two types, the horizontal axis and the vertical axis, the horizontal axis ones have to have a mechanism of some kind, such as a tailplane, to point them into the wind. Indeed, the windmills of 200 years ago had the first form of fully automatic control with a small tail rotor at right angles to the main rotor so that it was rotated whenever the main rotor was not pointing exactly at the wind and drove wheels on the ground to rotate the whole windmill. The old fashioned windmills had a relatively low tip speed of the same order as the wind speed and Fig. 4.15 shows how the power coefficients of various types of wind turbines depend on the velocity ratio of the tip speed to the wind speed. The high speed two-bladed type and the two or three bladed Darrieus vertical axis type give their peak power coefficients with a tip speed some five times the wind speed and the Darrieus has the further disadvantage that it is not self starting, although a number of other vertical axis types

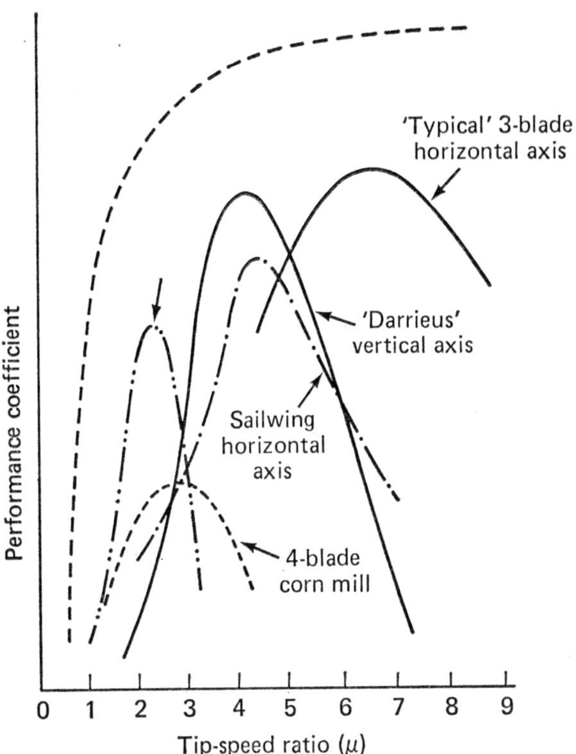

Fig. 4.15. *Cofficients of performance of wind turbines*

have been developed which are self starting, and much research is going on in this subject. The delta wing horizontal axis type being developed at the University of Calgary* is a very interesting new development which could be made quite cheaply.

The largest windmill ever built is that designed by Palmer Putnam on a 200 ft hill, Grandpa's Knob, in Vermont†. This has a 53 m diameter propeller and a tower height of 63 m, giving a propeller area of 220 m^2 and a power density of 560 W/m^2. The blade tip speed was 80 m/s and the generator 1225 kW. Full power was obtained at a wind

* J. A. C. Kentfield and D. H. Norrie, 'An axial flow wind turbine with delta wing blades', International Conference on Alternative Energy Sources, Miami, December 1977.
† D. M. Simmons, *Wind Power*, 1975 (Noyes Data Corporation, USA), p. 138.

speed of 15·3 m/s the blade angle was adjusted automatically and it was turned into the wind by having the propeller on the leeside of the tower. The machine was reckoned to have been a technical success but one of the blades broke off after it had been decided to terminate the tests because it was known to be over-stressed. It did, however, run for nearly four years, and gave enough data and experience to enable an improved version to be designed; this machine operated in the 1940s.

A great deal of work was done in Europe up to the 1960s, but work was stopped because at that time it was thought that nuclear energy would provide plenty of competitively cheap electricity.

The problems of covering a very large area of wind collection and concentrating wind as a suitable power supply to a generator have exercised the minds of many ingenious inventors. The simplest system is the use of a diffuser that converts all the kinetic energy of the arriving wind into a pressure rise downstream of the blades. In its simplest form this is a venturi-shaped collector with the wind turbine at the smallest diameter. It collects the wind over the whole of the frontal area. Augmentation factors of up to four have been produced; that is, the power output of the turbine is four times that which would be obtained with the same turbine without the diffuser. Such a diffuser has to be very large and long and has to be turned into the wind. The length has been reduced in one design by making the diffuser end in a louvre form.

Another very sophisticated proposal is the Tornado wind energy system*. In this, a tall vertical cylinder has tangental inlets so that whatever the direction of the wind a low pressure vortex is produced inside it. This, in turn, creates a low pressure up the axis which sucks air through a small wind turbine at the base; a velocity augmentation factor of up to eight is expected. It is also proposed to augment the suction by using a solar collector at the base, or even a device for burning coal, oil shale, or refuse, to create buoyancy in the vertical stack. The use of the cooling water from a power station has also been considered.

Another very ingenious system is the Madaras system. In this a series of spaced out wagons run on a circular track, each wagon carrying a tall, vertical cylinder, which is rotated by an electric motor; the wind acting on the rotating cylinder works like a sail and produces a force at right angles to both the direction of relative velocity of the wind and the wagons, and to the vector of the momentum of the cylinder along its axis. By reversing the direction of rotation twice on each circuit of the

* J. T. Yen, 'Harnessing the wind', *IEEE Spectrum*, March 1978, p. 42. M. B. Holland, 'Power from the wind' *Chart. Mech. Engr.* May 1978 p. 39.

wagon round the big track, the wagons are driven continuously in one direction, and power is taken off the system by the wheels driving generators. The aerodynamic lift coefficient of the rotating cylinders can reach a figure ten times that of ordinary air foils. It is estimated that with twenty cylinders, each 30 m tall and 7·5 m in diameter, 10 MW could be generated at a wind speed of 32 km/hr.

Since January 1978 the first Federally sponsored wind generator has been providing electric power to a community in the USA in New Mexico. This is a 200 kW two bladed conventional high speed wind turbine. The rotor starts automatically when wind exceeds 30 km/hr, and stops when the wind exceeds 64 km/hr. Rotor speed is maintained at a constant 40 rev/min by an automatic blade pitch mechanism. This is geared to a 1800 rev/min generator.

Many new proposals are now being put forward for large scale electricity generation. It has been pointed out that in Britain, if we placed a 10 kW generator every 50 m along the north/south 12·5 mile/hr isovent we should have 20 000 mills each saving 30 000 kWh/year, or 3·5 tons oil equivalent, saving 200 000 tons of oil, needed to produce 870 MWh/year. There are twenty million country dwellers in Britain, and if there was one machine to every five we should then have two million machines of 10 kW, with a hundred times the saving of oil. The basic problem of the use of wind is the storage problem, because the wind is very erratic. Electric storage batteries are certainly not feasible except in very small systems on an individual domestic scale for lighting purposes. Where electricity is being used at present for greenhouse heating or domestic space or water heating the windmill provides a very attractive way of saving electricity, especially in conjunction with night storage heaters. The possibility of operating a heat pump with a windmill, using solar energy to provide the low temperature heat input, is also a very interesting one for the future.

The simplest way of storing wind energy is to run windmills in conjunction with the National Grid, with controllable thermal power stations, so that the output of the thermal power stations is cut down as the wind feeds in.

A recent figure for an array of 1·5 MW units at a production cost of £280/kW, to be constructed in a shallow part of the southern North Sea, probably around the Wash, has been proposed, which could make a major contribution to Britain's total energy requirements. North Sea wind has an annual mean wind speed of 7·3 m/s, and it is possible to find 800 km^2 with a water depth less than 20 m; this could give 8 GW, that is 32 per cent of our average electricity requirements in Britain. The total capital cost, including a compression storage system in the exhausted natural gas strata under the North Sea, would be £505/kW, which compares with a 2·4 GW nuclear power plant ordered in New York State in July 1976 at a cost of £830/kW.

Thus, we can see wind as being useful in one of two ways; a large number of relatively small windmills of less than 100 kW, probably owned by a group of houses, or a large array of very large windmills greater than 1 MW, connected to the National Grid system. However, for those countries which have not yet got large electricity generating stations and have a high wind, either because they are near the coast or have high mountains, a large number of small windmills would appear to offer a much more interesting possibility for the future.

4.6.3 *Geothermal energy**

Naturally available geothermal energy in the form of geysers has been used for many decades in Iceland, Italy, and New Zealand. There are many places near major faulted areas in the world where local hot spots at temperatures of 200 °C occur at shallow depths. Such major faulted areas do not, however, occur in Britain. The rate of heat flow from the centre of the earth itself is far too slow to maintain a permanent heat source from underground, but it is possible to make use, for some twenty years or so, of existing hot spots, and there are regions of permeable rock, at depths of 2–3000 m, which can be made use of by passing water down one bore hole and getting it up another at about 80 °C; this has been done already to heat flats in Paris. It seems likely that we have, in Britain, permeable rocks suitable to provide the heat equivalent of a few million tons of coal a year by the year 2000, but this would only last for about twenty years.

There is not yet any practicable method of extracting the heat from dry rocks even though there are deposits of granite which are at adequate temperatures, and at the same sort of depths (about 2–300 m). This would require some kind of fragmentation of the rock at the base of the hole, and the use of a nuclear explosion for this purpose has been investigated in the USA. However, the dangers of radiation activity getting into the water brought to the surface would appear to be sufficient to make the process unattractive in human terms. There is also the possibility of upsetting the ground stability.

4.6.4 *Solar energy*

Solar energy is, of course, by far the most abundant renewable energy

* H. C. H. Armstead, 'The future of geothermal energy', *J. Inst. Fuel*, 1978, **51**, 109.

resource available to mankind, and, indeed, it is the primary source of all agricultural energy, and the solar energy of millions of years ago has resulted in all the stored fossil fuels. The sunshine falling on a 100 km square (1 M ha) in a favourable location (where it averages 300 W/m^2 over the whole time) is equivalent to 3400 MTCE/year, which is about half the present world energy consumption. The total fossil fuel chemical energy resources ($\sim 10^{13}$ TCE) are equivalent to the total solar energy falling on the earth in less than four weeks ($1{\cdot}5 \times 10^{10}$ TCE/hr). A 10 per cent efficient device for making use of solar energy could provide the whole of Britain's energy demand in 1973 using 7·8 per cent of the land and only 0·96 per cent of the land would be sufficient to provide our electricity demand.

Outside the atmosphere the radiation from the sun is at a rate of 1·35 kW/m^2 whereas in a fossil fuel or nuclear heat exchanger (e.g., boiler) it is possible to have a heat exchange rate of 50 kW/m^2. The maximum solar radiation obtained at sea level, near the equator, on a surface normal to the sunshine, is 1 kW/m^2 as a result of absorption in the atmosphere. Light cloud does not absorb very much radiation if one collects it from all directions, but it is no longer possible to use a focusing collector. The average solar energy input to a horizontal surface in Britain, day and night throughout the year, is 100 W/m^2, this is approximately one-third of the figure in the area of the Red Sea where 3400 TCE falls on 1 ha in a year. In Britain the horizontal surface input varies between 0·41 kWh/m^2–day (17 W/m^2) in December to 4·5 kWh/m^2–day (188 W/m^2) in June. The problem with the utilization of solar energy is even more a problem of storage, as the total solar energy in Britain varies very much from year to year. There may be many consecutive days with no sun and there is less solar energy in the winter when it is most wanted.

The growing of crops or wood is an excellent method of storing sunshine, and the theoretical efficiency of food synthesis in good land could be as high as 33 per cent. However, the actual efficiency of crops in temperate zones is 0·5–1 per cent and in the tropics 0·5–2·0 per cent. It varies from one crop to another, C4 crops, such as maize and sugar cane, being considerably more efficient than the C3 crops, such as the conventional cereals, wheat and barley. The utilization efficiency depends, also, on how much of the year the ground is covered with green leaves, and, in the case of growing trees for fuel, the period of root establishment corresponds to a period of low efficiency. Over the whole earth the process of photosynthesis only collects and stores less than 0·1 per cent of the sunshine. However, even with this low efficiency, the carbon fixed every year into stored energy by photosynthesis is more than ten times the total energy use of man.

The direct conversion of solar energy to chemical energy (or even to electrical energy) by carrying out the photosynthesis process *in vitro* is also being studied* since the process *in vivo* is so successful. However, the work is mainly in the stage of a fundamental study of the reactions, and is likely to involve a much higher capital cost per unit of converted energy than agricultural processes, so unless a much higher conversion efficiency can be achieved it is not likely to be a large contributor to our energy problem. Like the agricultural processes, it does store the sunshine accumulated over the year.

A great deal of research has been carried out on the direct conversion of the short-wave energy of sunshine into electricity by photo-voltaic cells. The original cells were silicon photocells with a conversion efficiency of some 10 or 12 per cent; GaAs cells have achieved a 23 per cent efficiency. However, the cost of such cells at present is of the order of $14/W, so they are far too expensive for anything except generating a few watts in remote areas. A great deal of research is going on to bring this cost down, and it is hoped that eventually it will come down to a competitive figure. The combination of the low energy input and the low efficiency requires a large area to be covered with the cells.

An extremely costly proposal† has been put forward for a geosynchronous satellite, which would rotate with the earth at such a height that it was always over a given part of the earth's surface and would only be shielded from the sun for 3/4 hr out of twenty-four. This satellite would carry 50 km^2 of solar cells on 'sails' which would always be pointing at the sun; the resulting electrical energy would be converted to radio frequency and beamed down to a receiver on the earth. The transmitting antenna would be 1 km diameter and the receiver 7 km diameter. This would produce about 10 000 MW. Such a scheme can only be described as a rich man's toy as the capital cost of putting it into orbit and maintaining it would be absurdly high. It is far more likely that photo-voltaic cells will be used in comparatively small installations, widely dispersed on the ground, giving power during the daytime only.

The other main use of solar energy is for direct heating purposes, either (i) by means of black flat plate collectors, usually covered with one layer of glass to reduce heat losses, or (ii) with focusing devices usually of the parabolic trough type with a tubular collector along the

* M. D. Archer, 'Photosynthesis *in vitro*', *J. Inst. Fuel*, June 1978, **51**, 100.

† P. E. Glaser, 'The case for solar energy', *Energy and Humanity*, 1974 (Peter Peregrinus, Stevenage).

axis. The flat plate collector has a low efficiency because it loses heat by radiation and convection through the same surface as that through which the sunshine comes. A glass cover is to some extent selective in that it transmits short-wave solar radiation almost perfectly but absorbs the long-wave infra-red corresponding to the low temperature radiation of the warm water. Much work has been done on more sophisticated absorbing surfaces to make them 'black' at the solar wavelength, which peaks in the visible, and 'white' for infra-red radiation. Figures as good as 80 per cent absorptivity of solar radiation and 20 per cent emissivity have been achieved. Another device which has been used is a honeycomb grill of fine tubes, long in relation to their diameter. When the black surface, with this grill in front of it, is pointed at the sun, the sun's rays fall directly on the black surface, but re-radiation is confined to the very small angle instead of taking place over a hemisphere. This system, however, loses the advantage of the conventional black plate collector which does not have to be pointed continually at the sun.

Detailed figures are available for the solar energy on surfaces facing at various angles and at various time of the year; for example, at Kew in Britain, a horizontal surface accepts about the same radiation in June as one with a 45° tilt facing south, the figure being about 17 MJ/m^2–day; a vertical surface facing south gives only 10 MJ/m^2–day in June. However, in December or January the vertical surface and the 45° tilt give a figure of 3·5 in these units, while the horizontal one gives only 2. On overcast days the horizontal surface gains over that on vertical surfaces facing south, because the radiation comes from all directions.

Turning to the focusing systems, these can be divided into tracking and non-tracking. A tracking system may have a conical parabolic or a cylindro-parabolic reflector, with the axis at right angles to the plane of movement of the sun, that is parallel to the axis of the earth at the equinox and 15° more horizontal at mid-summer, correspondingly more vertical at mid-winter. The axis thus has to be changed roughly every weeek. Tracking collectors not only give a concentration factor but they also enable the sun to be used from soon after sunrise to just before sunset, whereas non-tracking collectors, like flat plate collectors, only give effective sunshine use for about three hours on each side of noon. Various shapes of non-tracking collectors have been designed and tested; essentially, these are of cylindro-parabolic or even hemi-cylindrical with the axis running east and west horizontally. The central plane is that in which the sun rotates so it has to be changed once a week; the heat absorber is along the axis or focal line and is preferably black on the side facing the reflector and silvered on the opposite side. One very interesting experiment is that of Tabor who built cylindrical inflated transparent plastic sausages, with a boiler tube down the axis; the inside of this sausage is silvered half way round. With these it is possible to obtain a sufficient concentration factor to produce steam up to 200 °C. There would appear to be a very great need for small units, perhaps of sizes 1, 10, and 50 kW, to produce power in tropical countries which cannot afford oil. The design of reciprocating steam engines, using the latest developments in methods of production and materials, could be greatly improved as this subject has been relatively stagnant for the last fifty years.

The Power Tower* is being investigated as a large scale solar thermal electricity generator. In this, a boiler or furnace is placed at the top of a tower, which may be 500 m high. A large number of small mirrors cover the ground over an area of the order of the square of the tower height and each of these is moved automatically to reflect the sunshine that falls on it, on to the boiler at the top of the tower. This requires a fairly sophisticated control system and, of course, is subject to the usual problems of energy storage for night use, but, nevertheless, seems a valuable development in terms of the power obtained in relation to the capital cost of the installation. The French plan is to use a mixture of melted salts, heated from 400 °C to 500 °C in the solar furnace and stored hot, to provide steam from a heat exchanger as required.

Considerable thought has also been given to the question of solar cooling for buildings. One proposal is to use the conventional two-fluid refrigeration cycle worked by heat input, using a focusing collector and the ammonia–water system. Another proposal which has been installed is to have bags of water on the roof of a building which absorb the sunshine in the day, but radiate it at night. By putting a white cover over them in the daytime the sunshine is mostly reflected and then blackened surfaces at night give a good heat loss. The reverse process can be used in the winter when it is desired to use them to heat the building.

4.7 ENERGY ECONOMY

Many people have pointed out that *the consumer in a developed country could have exactly the same results from the use of energy with only half as much energy input*, provided it were economic to install known equipment for energy saving. By limiting energy wastage, for example, unnecessary packaging, advertising, built in obsolescence, and throwaway goods, a further reduction could be obtained.

* M. Barrere, 'Energy and environment', *Proc. Inst. Mech. Engrs.*, 1978, **192,** 197.

4.7.1 *Energy savings by the First Law of Thermodynamics*

In any heating process part of the energy input goes for the required process, part is lost through walls, and part is carried away in gases leaving the system. The thermal efficiency of such processes varies from 80 to 90 per cent for electric heating, to less than 20 per cent, for example, in the domestic open fire, or in the oil-fired open-hearth steelmaking furnace. By capital investment and good fuel technology this figure could be raised to over 50 per cent in all cases.

Heat losses through the walls could be halved in almost all cases by better insulation and better control of air infiltration, and the only reason we do not do this is that it is too expensive in capital cost. This applies to water heating processes, boilers, and furnaces, and to space heating.

In combustion heated processes the basic principles of fuel economy have been known for 100 years, but they require expensive instruments and careful control. Basically they are: (i) to obtain complete combustion, with a minimum of excess air, since excess air lowers the combustion temperature and therefore the fraction of the heat of combustion which is available for transfer at the working temperature, (ii) to have as good a heat transfer coefficient between the gases and the working surface as possible, so that with the available heating surface area the gases are cooled as much as possible.

In domestic heating the ventilation of air, especially with the open fire, may be as high as ten times the air required for comfortable ventilation. When open fires were replaced by electric radiators this unnecessary air was eliminated with enormous reduction in the heat requirements for comfort. However, when one considers the fact that the electricity is produced by throwing away three-quarters of the heat of the fuel in the power station and distribution system, the comparison becomes less attractive.

It has been estimated that something like 25 per cent of all energy consumption in Britain is used in domestic and office heating and hot water. The scope for saving at least half this energy, by better insulation, double glazing, better control of draughts and ventilation, the use of direct firing systems instead of consuming the fuel in electric power stations, and, ultimately, the processes which require no fossil fuel (solar heating and wind power), is quite clear.

4.7.2 *Energy savings by the application of the Second Law of Thermodynamics*

If one considers the cycle in which one burns a fuel, such as coal, in a boiler to make steam and then uses the steam in a turbine to produce rotary motion which drives a generator to produce electricity, one can look at the principal loss of energy in two ways. According to the First Law of Thermodynamics the boiler may be 80 or 90 per cent efficient, and the turbine less than 50 per cent. However, if one expresses the energy in terms of the Second Law of Thermodynamics, that is, in accordance with the fraction of it which could be turned into work in an ideal reversible thermodynamic cycle (discharging heat at ambient temperature) one arrives at an entirely different picture. Some thirty-five years ago I published an article* about the concept of 'virtue', which was defined as the fraction of the energy available in this way. The same concept has been reinvented in Germany and entitled 'exergy'. The 'virtue' or 'exergy' of a quantity of heat, Q, all at an absolute temperature T, is equal to $Q \times (1 - T_A/T)$ where T_A is the ambient temperature which is the lowest temperature at which heat can be exhausted. From this point of view it is easy to show that the greatest loss of virtue of the energy is in the boiler, viz. in the process of combustion, and in the transfer of energy from the combustion gases to the steam. The steam already has a much lower virtue than the unburned fuel because its heat is available at comparatively low temperature (mainly at the saturation vapour temperature). From the point of view of the Second Law of Thermodynamics, most of the energy is already degraded in the boiler and the most perfect steam engine cannot have an efficiency much better than that actually achieved. This is, of course, well known to all power-cycle engineers, who are striving to utilize more and more of the heat at a higher temperature, and was applied very clearly in the dual-cycle power generation system built by General Electric at Schenectady, where the boiler evaporates mercury, and the mercury, after giving up power, is condensed to give high pressure steam.

There are three ways in which this application of the Second Law of Thermodynamics, through the concept of virtue or exergy, can be applied to reduce substantially the primary fossil fuel consumption without any change in the resulting product. The first and most obvious of these is what is called 'Total Energy' ('Co-generation' in America). An example of this is Battersea power station, where pass-out steam from the power station is used to provide all the heat for a large block of flats. In this system the thermal efficiency of the electricity generation is reduced slightly because the exhaust temperature is higher, but the overall efficiency of heat utilization from the fuel can be over 70 per cent. The heat utilization has to be fairly close to the power station, although the larger the scale, the farther it is economically possible to carry the pass-out

* M. W. Thring, 'The virtue of energy, its meaning and practical significance', *J. Inst. Fuel*, 1943/44, **17**, 116.

steam. The Institute of Fuel held a Total Energy Conference in 1971, which it was shown conclusively that if, on the one hand, one uses fossil fuel to generate electricity and then uses more than half this electricity for domestic and office space and water heating, and, on the other hand, one uses a total energy system, it is possible to produce exactly the same result using 60 per cent less energy in the latter case. It is clear that from the point of view of the Second Law of Thermodynamics it is quite wrong to throw away two-thirds or even three-quarters of the energy in the fossil fuel in order to create electricity, which is a perfect thermodynamic fuel (that is, it has a virtue of 1, corresponding to $T=\infty$) and then use this electricity for raising temperatures by perhaps 30 °C, which corresponds to energy with a virtue factor of the order of 0·1. One can make out a strong case, in Britain at any rate, for not building any more electric power stations for some twenty years, but for gradually converting all the existing ones to coal-fired total-energy schemes, and, at the same time, installing fluidized bed combustion systems and very large, improved, gas-cleaning and sulphur dioxide removal systems.

The second method of applying the Second Law of Thermodynamics is the heat pump. When you explain to someone who has not understood the law that it is possible to produce three or four units of heat in hot water, or in space heating, by using 1 unit of electricity, he is liable to accuse you of being unscientific. However, in 1850, Lord Kelvin pointed out that this is just what one does with a refrigerator, and much work has been done on heat pumps ever since. The performance factor of a heat pump is defined as the heat given out by the system at the higher temperature, divided by the mechanical work done to operate the cycle. The heat pump is necessarily expensive, and is a good example of the fact that saving energy requires capital cost. However, there is little doubt that in cases where it is essential to use electricity for heating, it should, in the future, always be done with heat pumps rather than directly. It might even be possible, with improved heat pump designs, to produce a heat output equal to the total heat input required at the power station to generate the electricity used. The other main problem with a heat pump is the source of the low temperature heat; four systems have been tried. The most obvious is where a river is available and can provide the source. Air can be used, pumped through the system, but this gives difficulties of frosting. More recently, the idea of a solar heater with a large warm water store acting as the low temperature heat source is being investigated. This obviously depends on the availability of sufficient sunshine at not too infrequent periods throughout the winter. Recent work has also gone into use of the ground, which can provide a sufficient source of heat throughout the whole of the winter provided the pipes cover a very wide area and are buried at a reasonable depth. One important point about heat pumps is that the heat output must be at as low a temperature as possible, consistent with the use of the warm air or warm water, in order to give a reasonable performance factor.

Other ideas being investigated are: (i) the recovery of all the heat in warm water discharged from bath and kitchen sink for the heat source, (ii) a heat exchanger between incoming cold air and outgoing warm air. None of these ideas are economic as yet, owing to the capital cost of the installation. If air conditioning is required in the summer, a combined heat pump for the winter and cooler for the summer becomes more attractive because there is the capital cost of the air conditioner to be set against that of the heat pump. Another scheme that has been tested and is obviously helpful to the economy is the combination of an ice rink and a swimming pool using a heat pump, with the ice rink as the heat source and the swimming pool as the heat sink.

It has been estimated that 2·5 per cent of the total energy consumption of the UK is used for evaporating water in drying processes. The largest users are the paper and board industries, and the brick-making industries. The sugar industry has worked on the reduction of its primary energy for evaporation for many years by using multiple effect evaporation, but this is not feasible when drying solids, and some 9 M tons of water are removed every year from these solids by direct heating. There are two ways in which the heat pump principle can be used for drying: (i) where the drying is at a fairly low temperature, as in the paper and board industry, the heat pump based on an air compression cycle can be used, (ii) higher temperature drying could be carried out by the vapour compression cycle, in which the latent heat of the water is recycled by condensing it at a higher pressure to transfer the heat through a wall to evaporate the moisture at lower pressure. The water leaves as liquid, and the input to the system is the energy for compression.

The third method of applying the Second Law of Thermodynamics to fuel economy is to use the energy at different temperatures for different purposes. The use of waste heat boilers on furnaces, gas turbines and diesel engines is a well established technique, which is, however, rarely applied, because the capital required is too large in relation to the value of the fuel saved. The other main example is the use of recuperators and regenerators to cycle the heat below the working temperature of the process back into the process by transferring it to the air for combustion. This raises the combustion temperature and thus makes much more of the heat available above the work temperature of the furnace. Again, this has been used in industry for many years, particularly in those industries such as the open hearth steel melting furnace, where the required temperatures could not be obtained without the pre-heat. The continuous regenerator, or 'heat wheel', is a method which has been receiving increasing interest of late. As in all pre-heat systems it is essential to keep the combustion gases and the pre-heated air as hermetically apart as possible, since carbon dioxide and water in the combustion air lower the combustion temperature very substantially. A recent modern idea is the use of a battery of heat pipes to carry the heat from the channel with the exhaust gases to the channel with the incoming air. It is still necessary, of course, to increase the heat transfer on the outside of the heat pipes to the two gases by means of fins, since this becomes the major bottleneck in the system. However, this does make the avoidance of leakage much simpler.

4.7.3 *Energy economy by the elimination of waste*

Because fuel has been cheap in the developed countries we have based the whole of the Industrial Revolution on an almost total disregard for the exhaustion of the limited fossil fuel supplies. The use of unnecessary plastic packaging, throw-away cans and bottles, the use of electricity for advertising, the built-in obsolescence of, and annual fashion changes to, consumer goods such as cars, are all examples of the use of energy in ways which are of *no benefit whatever to the consumer*. The food industry uses considerably more energy in 'upgrading' food than is used to grow it. Much of this upgrading consists of refining processes, in which the most valuable constituents are removed and then sold back to the consumer, for example, as vitamin tablets and bran. It is generally agreed that the untreated foods would be much more healthy. A great deal of the energy required to obtain metals from low grade ores and all the energy required to fix nitrogen and to mine phosphates for fertilizers could be saved by recycling processes.

A final example of waste is the flaring of natural gas at remote, inaccessible oil wells, and the flares at oil refineries and coke oven batteries. In Britain we throw away in our dustbins some 5–6 M tons of paper a year. This is the principal source of calorific value in refuse (the contribution from plastics is less, but also substantial). We have not so far found any system for recycling all this paper, although it can be used for making hardboard and egg boxes, or cleaned of printing ink and reused as paper*. Work is going on on the incineration or pyrolysis of refuse, and a number of plants have been operating for many years, but here again the fuel saving has not, so far, made the capital expenditure worthwhile.

* This book is printed on paper so recycled.

4.7.4 *Conclusions on energy economy*

Many studies of energy economy have been made and estimates for the energy that could be saved by investment of capital in already well established fuel saving methods range from 20 to 50 per cent. If to this is added the elimination of the wasteful uses of energy which are of no benefit to the consumer, discussed in section 4.7.3 there is little doubt that *all the rich countries could reduce their energy consumption to less than half the present figure with no change whatever in life style.*

4.8. A WORLD ENERGY STRATEGY

4.8.1 *The engineer's responsibility for energy*

When the politicians and the economists make forward projections of energy supply and use, they are forced to make them in terms of well established existing machinery*. It is only the engineer who can see what new machines are possible within the limits of the known laws of science and the known or estimated resources of the earth. Moreover it takes more than thirty years to develop a new machine to the point where it can make a world wide impact on the energy situation. Thus, the engineer has a unique responsibility for looking forward to the future and establishing a world energy strategy which can give our descendants a decent life in the next century. In this task the engineer has three groups of opponents. The first and most obvious group is the politicians and the civil servants, whose job it is to preserve the present situation and win the next election by giving, or promising, people more and more and never asking for sacrifices. This forces them to take, always, a short term view of the energy problem, and to maintain large existing projects long after the original promise of these projects has been totally disproved.

The second group of opponents are the economists and accountants who calculate everything in terms of next year's profits and the rapid repayment of capital investment. This forces them to insist on fuel being kept so cheap that it does not pay to economize on it. An outstanding example is British North Sea gas, which will be essentially used up in twenty-five years. The only reason for this unkindness to our grandchildren is that the economists insist that the capital investment in the field be repaid as quickly as possible.

The third group of opponents are the most difficult of all to argue

* The research and development on nuclear energy is an exception, but here the military demand for nuclear explosives has been the deciding factor.

against because they are themselves engineers and scientists. These are the people who have a vested interest in policies which can fairly easily be shown to be disastrous on a long term basis. Their careers depend on the successful development of one particular source of energy, such as nuclear energy, or the steadily increasing sales of one particular form of energy, such as electricity, gas, or oil. Since these people have much more direct access to the necessary scientific facts and figures it is fairly easy for them to accuse those engineers who try to take a longer term view of emotional judgement and of not knowing the correct facts. As the whole theme of this book is that we all have to train the emotional ability of our conscience to see what is right for humanity, I am not ashamed of being accused of emotionalism. However, it is necessary to answer these arguments as logically and scientifically as possible, taking as long term and as broad a view as possible.

For example, it is relatively easy to design nuclear power stations to make the risks of radioactive escape negligibly small, and even, ultimately, to find a safe way of disposing of all the waste. However, when one takes account of the facts that:

(1) human beings have never managed to escape periodic wars and violent crimes, using the latest weapons, nor are they immune to errors and misjudgements,

(2) we are unable to see any way of dealing with derelict power stations,

(3) they could never be built in sufficient quantities to satisfy all the needs of the underdeveloped countries,

logic points in a very different direction. Similarly when one applies the Second Law of Thermodynamics to the generation of electricity, the use of electricity for space heating and water heating becomes totally illogical.

Both the politicians and the economists are gullible in engineering terms. For example, many papers have been written about hydrogen replacing oil and natural gas for transport and other energy purposes*. The engineer knows that, not only is hydrogen a very much less convenient and safe fuel, but also that it requires immense capital cost to produce it, whether it is produced by the generation of electricity in nuclear power stations followed by the electrolysis of water, or produced by *in vitro* solar systems. At one stage in the honeymoon of nuclear energy, plans

* J. K. Dawson, 'The prospects for hydrogen as a fuel in the United Kingdom', *Atom*, May 1974, No. 211, p. 93.

were even worked out for producing hydrocarbons by electricity from the carbon in chalk and the hydrogen in water!

4.8.2 *Long term versus short term decisions*

Professor Carol Wilson has written*

> When a problem has been ignored as long as we have ignored or misjudged the energy problem, the short term requirements may not mesh with the medium or long term, and so it is in this case. We have to fix the roofing and build new houses at the same time.

The consequences of the short term decisions over the last thirty years in Britain stem from the post-war assumptions that Britain could have plenty of cheap oil for a long time and that, ultimately, nuclear electricity would be so cheap that energy would not have to be conserved, even when the oil ran out. Some of these consequences can be listed as follows.

(1) We reduced our coal output to about half what it was just after the Second World War by closing the less economic pits.

(2) We scrapped most of the work on fuel economy, which had run at a very high level in the few years after the war, when the first edition of the *Efficient Use of Fuel* was published (1950).

(3) We closed down many of our railways and built more and more motorways when railways are the only form of transport which can run just as well on coal as on oil; and the motorways will be spoiling good agricultural land long after the oil has run out.

(4) We built many power stations for oil firing and even considered natural gas firing.

(5) We built many houses and flats without chimneys so that they are forced to use electricity for space and water heating.

4.8.3 *Is there a solution to the world energy problem?*

Clearly the engineer cannot solve the world energy problem alone because this requires sacrifices, efforts, and long term political decisions. Nevertheless, the engineer can show whether there is a feasible solution, and I shall now try to answer this question in the affirmative and establish the necessary conditions.

The first point to be made is that *technology has shrunk the world to the*

* C. L. Wilson, 'World energy prospects for the year 2000', *Proc. R. Soc., A*, 1977, **358,** 121–139.

point where no one country can consider its energy problems in isolation. In the first place millions of tons of oil, iron ore, coal, phosphorus and food are moving round the world in large cargo ships all the time. Japan is almost entirely dependent for its industrial growth on imported raw materials and exported manufactured goods. Secondly, people can travel halfway round the world in aeroplanes in 24 hr and communicate halfway round the world by satellite in milliseconds. Thirdly, and most menacingly, the fantastic development of weapons of war has given us inter-continental ballistic missiles which can go halfway round the world in less than an hour carrying nuclear warheads, the radioactivity from which would spread over the whole hemisphere. Professor B. F. Feld says*

> The United States and the Soviet Union alone—the two so-called superpowers—together possess more than 50 000 nuclear bombs having explosive power ranging from a fraction of the destructive force of the Hiroshima and Nagasaki weapons up to bombs of more than a thousand times the explosive intensity of those horrible but primitive instruments of death.
>
> An all-out nuclear war today between the two superpowers would annihilate the populations and destroy the treasure of both nations—and this is irrefutably true irrespective of any measures, offensive, protective or otherwise, that either of the two giants could take. Both countries would, for all practical purposes, cease to exist. But that is far from the whole story.
>
> The lethal radioactivity, carried by the prevailing winds well beyond the borders of the victim nations, would take a toll of many tens of millions of innocents throughout the northern hemisphere. And yet the devastation would not be confined to the countries of the northern hemisphere alone. Radioactivity, spread throughout the globe by stratospheric air currents, would eventually rain down on all peoples, everywhere, from the arctic tundra to the vast continent of the antarctic. Not intense enough to be lethal to all—it is true—but enough to claim a vast toll of radiation-connected diseases in this generation, and incalculable damage to future generations in hereditary malformity and lethal mutations. Not enough is known about the effects of such a large genetic load to be able to say, at this time, whether the survival of the species would thus be placed at risk. But there is no doubt that the sufferings would be fantastic and the risk very great.
>
> Yet what, to me at least, is even more frightening is that the current and anticipated rates of accumulation of materials capable of being made into nuclear weapons—through peaceful nuclear fission energy programmes in addition to military programmes—is such that by the end of this century, enough weapons material will be available, and perhaps enough weapons as well, so

* B. F. Feld, speech at the Plenary Session of the UN Special Session on Disarmament, 12 June 1978, *Pugwash Newsletter*, July 1978, **16**, 2–3.

that their use in an all out nuclear war would guarantee the elimination of the human species from the entire earth.

As soon as one takes a long term view, therefore, one has to consider the whole of humanity, that is the needs of the 8000 M people who will be on the earth early in the next century. It is not possible for my grandchildren in Britain to enjoy a peaceful life with enough of all the necessities unless the whole of these 8000 M people enjoy the same. Thring's economic principle states that '*Whatever is right for my grandchildren is always uneconomic now and almost always impolitic*'.

The question of how to level off the growth of the world's population has already been discussed in Chapter 3, and it was concluded there that the only humane way of doing this is to give a decent standard of living and a decent education to all the people of the world, and this must be achieved within thirty or forty years from now, before the problem reaches the point at which world-wide disaster and pestilences occur.

It has also been pointed out in Chapter 3 that to avoid the Third World War or total breakdown of law and order, with banditry and freedom to murder one's neighbour in self defence, we must essentially eliminate the gap in standard of living between countries and between groups of people in the same country, distinguished by one or another kind of 'football shirt'.

It follows that *the only machines and processes which will be viable* if we are to have a peaceful twenty-first century *are those which could be made available to all humanity*, everything else are 'rich men's toys'. The capital and energy investment and raw materials required to make large central power stations and the resulting electricity distribution network are such that they could not possibly be provided for all 8000 M people all over the world.

It is probable that small, widely distributed, electricity generating units, ranging from 1 to 500 kW, spread all over the earth, will produce the most feasible solution, although there may be bigger power stations up to perhaps 200 MW in the larger towns. Some industries, such as steel making, which really do show enormous economies of scale, may continue to be located in large central plants at appropriate sites in the world, particularly near to good iron ore or coking coal deposits. These will be mainly for making steel from ore, and scrap will be re-melted locally, in much smaller plants. Fuel-fired counter-flow melting furnaces can give very high efficiencies for scrap remelting.

Now we come to the real crux of the argument. At present the rich countries are consuming 5 or 6 TCE/person–year, and the poor countries about 0·5; the rich countries are even planning to increase still further with low growth of 2 per cent/year or high growth of 4 per cent/year, and anyone who talks of zero growth is regarded as a visionary. If we are to eliminate the gap between the rich and the poor countries by bringing the poor up to even the present figure of the rich countries, we should have 8000 M people consuming an average of 6 TCE/person–year, that is, more than six times the amount of energy consumed at present; some 48 000 MTCE/year. This is not a feasible scenario. We cannot provide this much energy in any way. Nuclear fuel would require too much capital cost for the power stations. The reasons against fossil fuel are the limited supplies of these fuels, cost of extracting them on this scale, and also the very real danger of carbon dioxide rising in the earth's atmosphere to a point where the 'greenhouse effect' melts the polar ice caps. Renewable energy cannot be provided at this level, again because of the capital cost of the plant.

One is, therefore forced inevitably to the conclusion that the only possibility that leads to a feasible and satisfactory civilization in the next century, is for the rich countries to reduce their per capita energy consumption to a figure about equal to the overall world average figure. At the present time that is about 2 TCE/person–year. Thus, the only

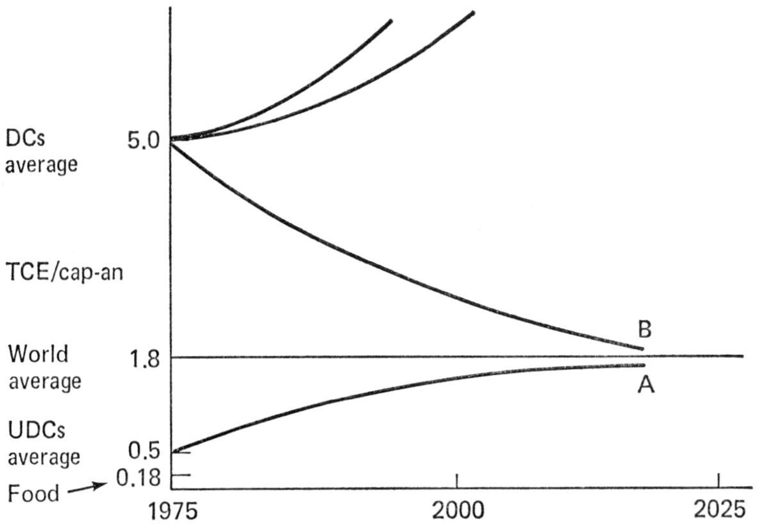

Fig. 4.16. *Present and future energy consumption/person–year*

future that makes sense is *a negative growth scenario, in which the energy consumption per capita in the rich countries goes down by 3–4 per cent each year for the next thirty years.* In the USA the reduction figure would have to be about twice this.

Figure 4.16 illustrates a feasible world energy policy for the future in terms of TCE/person–year. Any attempt to have growth, or even to maintain the figure constant, in the developed countries must lead to disaster within the next thirty years. The following quotation is from the preface by Professor Sir Martin Ryle to the booklet *An Alternative Energy Strategy for the United Kingdom*.*

> We have become accustomed to believing that our standard of living and full employment both depend on an ever increasing consumption of energy—to heat our homes, to provide more transport and especially to produce increasing quantities of manufactured articles which consume both energy and other mineral resources, some of which have limited reserves. If we continue in this belief, the energy shortage initiated by the exhaustion of the world's oil reserves will, by the end of the century, forcibly curtail the present trends in the worst crisis civilization has yet experienced. The time-scale is so short that it is not clear that any viable nuclear programme can avoid the energy problem, while providing no solution to the exhaustion of other resources.

4.8.4 *Premium and non-premium fuels*

It has been demonstrated that, even on the most optimistic estimates, oil and natural gas are premium fuels because, on the one hand, we are using them up at a rate such that they will only last a few decades and, on the other hand, they are uniquely suitable as a cheap and convenient source of energy for certain essential purposes, such as transport, petrochemicals, and agriculture. It is, therefore, a criminal neglect of the interest of our grandchildren to waste the premium fuels now by using them for purposes for which coal is suitable, and expect them to have to use a very expensive substitute or a very much less convenient alternative for the premium purposes. It will take 1·5–3 tons of coal to make 1 ton of oil substitute, and the capital cost of the plant will be very high indeed, and this has to be added on to the cost of mining the coal. If the use of these two premium fuels is reduced by about 8 per cent/year every year for the next thirty years in the developed countries, bringing their use down to one-tenth of what it is at present, then it will be possible for the underdeveloped countries to have a fair share, and for the exhaustion of the premium fuels to be postponed for several more decades.

* *An Alternative Energy Strategy for the United Kingdom*, 1977 (National Centre for Alternative Technology, Machynlleth, Wales).

Premium fuels must be used for premium purposes only, and with very great economy even for these purposes. Electricity must be regarded as the premium fuel par excellence because of the low overall thermal efficiency of its generation and the high capital cost of the generation and distribution equipment. It is a sin to our descendants to use it for any low grade purposes, and its use is only justified on a long term basis for lighting, power, electrolysis, and very high temperature processes (over 1600 °C), and even for these it must be used with great economy.

4.8.5 *The ultimate energy supply*

The non-premium fuels are coal, various forms of waste such as colliery spoil and domestic refuse, and all the renewable fuels. However, even coal will be exhausted eventually, and we must therefore regard all fossil fuels as capital, and ensure that enough of it is spent in installing equipment to use renewable energy sources so that by the time the fossil fuels are all used up these sources can provide the whole of humanity with a standard of living corresponding to about 2 TCE/person–year. We should certainly be doing very much more research, at the multiple prototype level, on small solar heaters, wind power units, on the provision of lighting by means of very much lower wattage systems, on heat pumps, and, above all, on biomass. As it is only the rich countries which have resources to spare for solving the world's problems, they have to find means of making it financially worthwhile for the small and domestic consumer to install systems which are very much less dependent on fossil fuels. We have to find fiscal systems in which, on the one hand, fuel saving systems become much cheaper to install, and on the other hand, fossil fuels become more expensive. The engineer has to persuade the politicians that advertising cars on the basis of how quickly they reach 100 miles/hr is a much bigger offence against our descendants than smoking is against ourselves.

CHAPTER 5

Transport and communications in the twenty-first century

5.1 WHAT TRANSPORT DO WE REALLY WANT?

It is in the field of transport that the engineer's contribution to civilization is most clearly marked as shown by the difference between Concorde and the space rocket on the one hand, and 'the ploughman homeward plods his weary way', on the other. It is, however, in the field of transport that it is above all necessary for the engineer to be clear as to what human needs he is satisfying. Concorde is the result of the argument that speed has always increased and must go on increasing as an automatic process, and that the engineer should concentrate his efforts on a very few rich people who find it convenient to commute across the Atlantic in a time considerably less than the difference of solar time. Many engineers are devoting their attention to increased speeds, and we still see motor cars advertised on the basis of how quickly they can reach 100 miles/hr when there is a legal speed limit in Britain of 70 miles/hr. At the same time we see all the great cities of the world becoming totally clogged with big lorries and private cars, so that often they move at no more than a walking pace for a mile or more.

Bouladon*† produced a figure, reproduced in Fig. 5.1, in which he divided transport into five areas, suggesting that the optimum way of moving a distance up to one-third of a mile was on foot and that in the range from one-third of a mile to three miles there was a 'transport gap', although the cable car, the horse, the bicycle, and the bus come in the centre of this region, and the underground train at the top of it. The third region, from three miles to thirty miles, he said was the principal region for the private car, although the train and scooter also come into this region. The fourth region is his second transport gap between thirty and 300 miles, and with optimum speeds of over 100 miles/hr. Here the only present systems are the helicopter, the high speed train or tracked air-borne vehicle, and the short take-off aircraft. Above 300 miles he suggested

* G. Bouladon, 'The transport gaps', *Science Journal*, April 1967, p. 41.
† G. Bouladon, 'Transport', *Science Journal*, October 1967, p. 93.

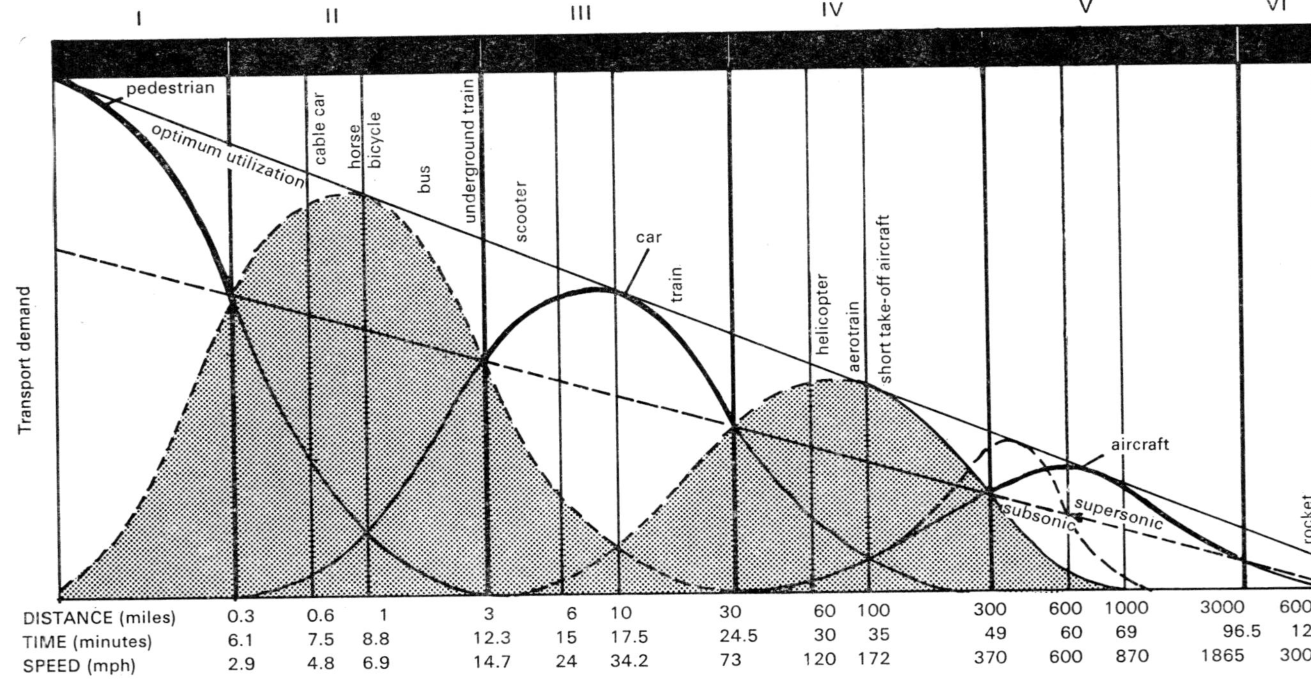

Fig. 5.1. *Transport gaps* (*shaded*) *can be identified by plotting demand for transport* (*vertical axis*) *against speed and optimum range of existing systems. The range is divided into six areas, of which I, III, and V are catered for by pedestrian, car, and air transport. Notable gaps occur in areas II and IV. Demand for short journeys is much greater than for long ones.*

the normal subsonic and supersonic aircraft fill the needs. As a result of this analysis he proposed a 'continuous transport' system in which passengers can get on and off at stations where the integrator accelerates them alongside a continuous transporter moving at 20 miles/hr. This would be a considerable improvement over the underground train as it would carry up to ten times as many people, (since there are no gaps between trains) and the fuel consumption would be very much less, while the time for most journeys would also be less because one would not have to stop at the intermediate stations. Many other proposals have been made for filling what Bouladon calls the lower transport gap, where the demand, from city commuters and visitors, is enormous. However, although underground trains are being installed rapidly in many cities (before the war there were about twenty, now there are about fifty, and there may be eighty in ten or fifteen years from now*) these are all of conventional type with rails and separately propelled trains, usually rising above ground in the more remote parts of the city where land is not so expensive. So far no city has had the courage and the finance to install one of these continuous transport systems, in spite of its advantages, and it is unlikely that it will be done because of the enormous capital investment.

When we look to the future it is certainly unlikely that the present trend will continue, that is, that cities will get larger and larger and more and more crowded with private cars. There is already a move out of big cities, but these are only minor palliatives to the problem. As far as the developed countries are concerned the problem must be considered in relation to the problem to be discussed in Chapter 7, namely, how to provide a worthwhile job for everyone, which undoubtedly means that the number of people employed in large centralized factories and the large centralized head offices of vast industrial complexes will go down very considerably. It also means that the present trend to automate public transport systems to reduce the number of people employed, and then pay them a dole for doing nothing, will be regarded as nonsense in the future. The elimination of bus conductors and ticket collectors is an example of what public transport systems are forced to do, to the considerable disadvantage of the travelling public, and to the detriment of the employment figures. A study of the unemployment problem, in Chapter 7, also indicates that there will be many more people working on the land in the developed countries, in the future and it may pay people to

* P. E. Garbutt (Chief Secretary, London Transport), 'Transport in cities', *Jl R. Soc. Arts*, June 1978, p. 422.

commute out of the towns to the land in special buses at certain times of the year, when more work on the land is required.

One fact which shows up clearly is that the number of private cars in towns ruins the convenience of bus systems, and it will certainly be necessary to eliminate, to a very large extent, private transport within towns, and replace it with very much improved public transport of all the existing kinds. *The primary job for the engineer* is no longer the increase of speed and automation, but *is the achievement of fuel economy, noise and pollution reduction, and the improvement of convenience, comfort, and reliability in all public transport systems.*

Proposals for better integration of systems of transport such as that which I put forward in my book *Man, Machines and Tomorrow**, may also give greater flexibility to the transport system and greater convenience to the public. Unfortunately, they do not involve exciting engineering research and development in the way that increasing aircraft speeds does. The proposal was to use individually-owned luggage-trolleys-cum-seats, which could be wheeled to a self-drive taxi rank, a bus stop, a railway station or an airport and locked on to the system for the public transport journey. The public transport system would simply consist of buses or trains or aircraft without seats, but with platforms on to which these units could be locked.

In the next sections of this chapter I shall consider the problems of improving the performance of existing transport systems in these directions. It will probably be too expensive to restore country railways in Britain where they have been destroyed, but mini-buses or multi-passenger taxis will have to be introduced.

5.2 ENERGY USE IN TRANSPORT

Table 5.1, taken from the paper by G. Leach†, shows that all transport took 32 per cent of the energy in the USA in 1960, while this figure had fallen to 28 per cent in 1969. In the European OECD countries, the corresponding figures were just over 16 per cent for both periods, but whereas the road transport was a fairly steady 80–85 per cent of all transport in the USA, the corresponding figures rose from 60 to 86 per cent in OECD Europe. In both cases the passenger car took around 70 per cent of the total used in road transport. Air transport was 3–4 per cent of the total energy consumption in the USA, and rose from

* M. W. Thring, *Man, Machines and Tomorrow*, 1974 (Routledge and Kegan Paul, London).
† G. Leach, paper in M. W. Thring and R. J. Crookes (Eds.), *Energy and Humanity*, 1974 (Peter Peregrinus, Stevenage), p. 109.

TABLE 5.1 TRANSPORT'S SHARE OF NET ENERGY CONSUMPTION: USA AND OECD EUROPE

	USA		*OECD*	*Europe*
	1960	*1969*	*1960*	*1969*
Total net energy consumption*†	8865	14747	5594	8876
Consumption all transport:	2815	4111	898	1447
as percentage of total	31·8	27·9	16·1	16·3
Consumption road transport:	2369	3356	535	1240
as percentage of total	26·7	22·8	9·6	14·0
as percentage of all transport	84·2	81·6	59·6	85·7
Consumption by passenger car:				
as percentage of total		16·1		9·4
as percentage of road transport		70·7		67·1
Consumption by air transport:	302	596	55	138
as percentage of total	3·4	4·0	1·0	1·6
as percentage of all transport	10·7	14·5	6·1	9·5

Source: OECD energy statistics 1955–69, pp. 188–205 and 111–129.
(All data from source given except those for passenger car, which come from various sources.)
* 'Net consumption' is total useful energy consumed; i.e. excluding energy losses in electricity generation and fuels used for non-energy purposes.
† Units: 10^9 kWh equivalent.

1 to 1·6 per cent in OECD Europe. A more recent paper by Bouladon* suggests that the private car consumes 60 per cent of the world's oil energy, and that if consumption by private car could be reduced by 50 per cent, the oil problem would be solved, at least for twenty-five years. This does not take account of the needs of the underdeveloped countries. During the five years up to 1974 the average specific consumption of the American car fell from 17·5 to 15·8 miles per Imperial gallon. The average weight of the American car was in the region of 2 tons, and each year it consumed its own weight in fuel. The American car manufacturers were still pushing at the time for heavier and more automatic vehicles, which used up more and more energy. It is only since the oil crisis of 1974 that the attempt has been made to call a halt to this trend.

The Karman–Gabrielli plot (Fig. 5.2†) takes account of the fact that the air resistance is proportional to the square of the velocity by plotting

* G. Bouladon, 'Towards a better utilization of energy in transport', Royal Aeronautical Society, Spring Convention, 1974.
† *Loc. cit.* D. Howe, Contribution to discussion.

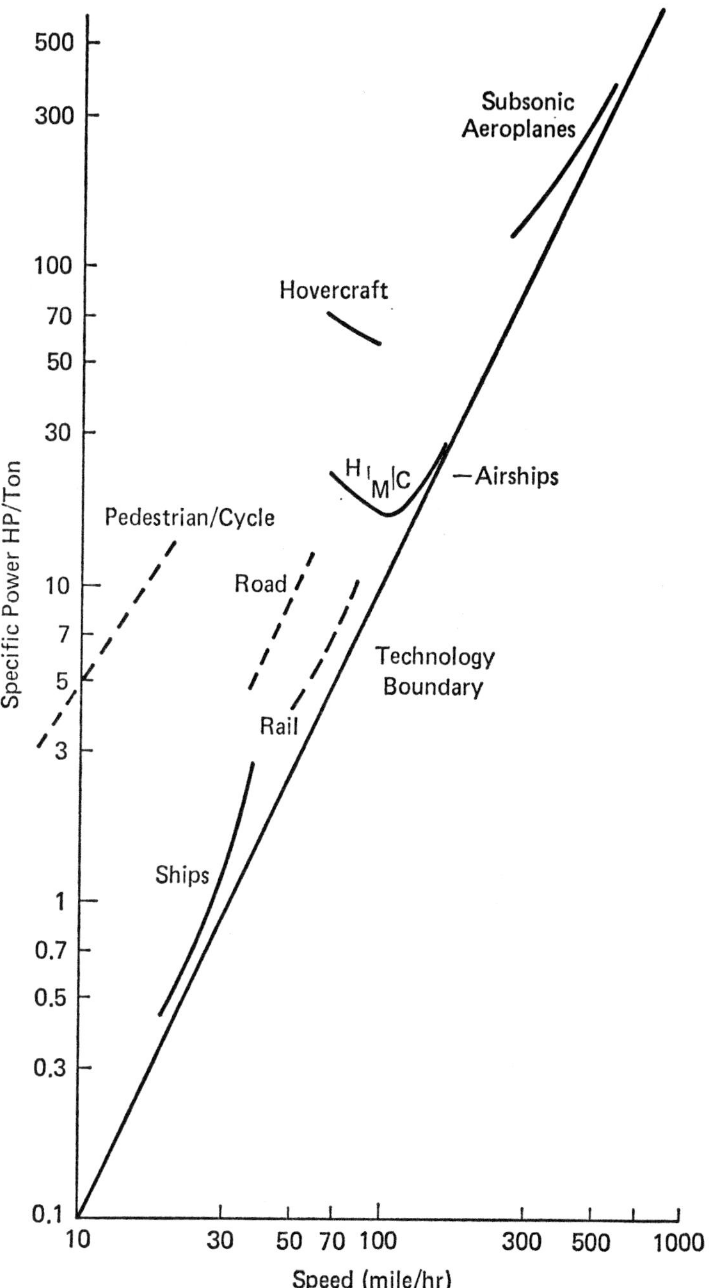

Fig. 5.2. *Karman–Gabrielli plot*

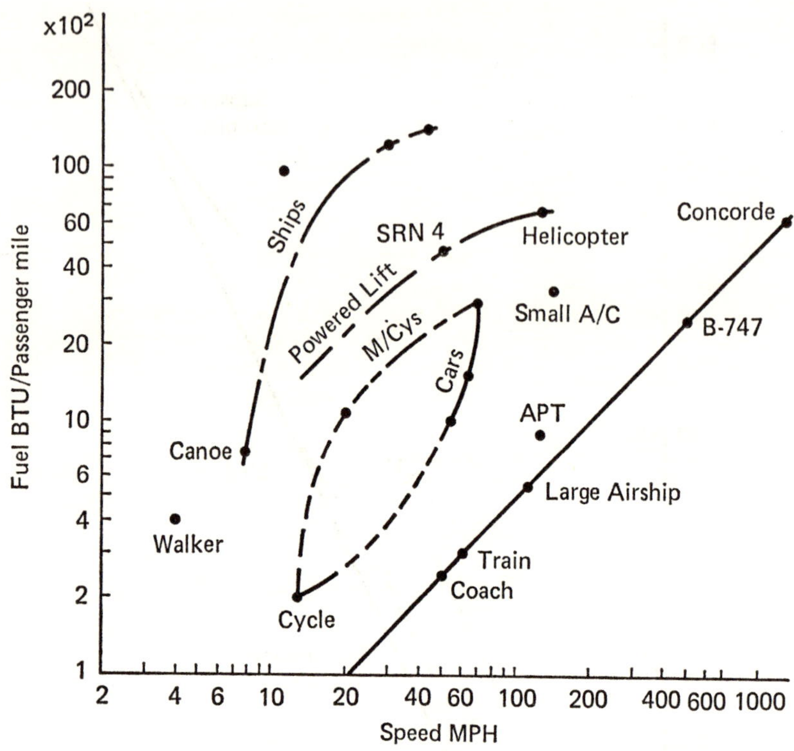

Fig. 5.3. *Efficiency of fuel usage*

the specific power in horse power per ton (vehicle plus payload) against the velocity on a log/log plot, with a technological boundary corresponding to this square law*. On this basis, of course, the ship comes out as having by far the lowest power per ton, illustrated by the barges, where a single horse could draw 100 tons, but the velocities are correspondingly low. The hovercraft obtains much higher velocities, but at the expense of a very large increase in energy. Rail comes much closer to the technology boundary than road because on road vehicles a lot of power is absorbed in the tyres, and there is also a greater braking loss due to more

* This plot assumes that air or water resistance is the fundamental energy limitation, with a limiting lift/drag ratio for aircraft. For ground vehicles tyre hysteresis and other factors are of comparable importance—racing cyclists use very thin, hard tyres.

frequent stopping. Figure 5.3 produces rather different relative order because here the fuel energy per passenger mile is plotted against the speed, similarly on a log/log plot. Here the bicycle shows up much better because the cycle weighs considerably less than the rider does; ships now show up considerably worse because they are very heavy in relation to the passengers they carry. A fully loaded train is still on the technology limit because it has relatively low air friction per passenger, and this is the main energy loss at high speeds. The fixed wing aircraft, when fully loaded, comes close to the limit set by air friction. These figures show clearly that speed costs a lot in terms of energy, but they also show that even after the speed is decided there is still considerable variation between different methods of transport in terms of minimum energy consumption.

This applies particularly to the car. Leach* pointed out that in 1974 the average car in the USA consumed 2 tons of gasoline every year, while in the rest of the world it was 1 ton; lorries and buses each consume 4 tons a year. Earl Cook† gives an analysis of why the American motor car is so inefficient. He lists eight factors.

(1) The car is about twice as heavy as it needs to be.

(2) The deformation of soft tyres for comfort consumes about 30 per cent of the energy.

(3) Trips less than 5 miles mean that the engine is running cold and inefficiently.

(4) Radio and head lamps consume considerable electrical energy [one might add the radiator fan which consumes several horse power at high speeds, when it is quite unnecessary. This can be avoided if the fan is electrically driven with automatic control].

(5) It is designed for fast acceleration so that it cruises at about one-third of its optimum power output.

(6) High exhaust temperature.

(7) More than half the time one has a speed different from the optimum.

(8) Normally far less than optimum load. A car carrying four passengers

* *Energy and Humanity*, p. 110.
† Earl Cook in J. Lenihan and W. W. Fletcher (Eds), *Energy Resources and the Environment*, 1975 (Blackie, Glasgow).

uses hardly more than a quarter as much fuel per passenger mile than if only the driver is there.

Cook points out that the fuel used by private vehicles in the USA is equivalent to the entire energy consumption of Japan, which is the world's third largest energy consumer. The increasing need for the USA to import oil is the principal cause of its balance of payment problem.

One must also take into account the energy costs of manufacturing a car, since modern cars last considerably less time than those constructed before the Second World War. Leach quotes the analysis by Professor Berry of Chicago University showing that the energy costs of manufacturing a large American car, when one adds up everything, including the transport of raw materials and the smelting of the metals, amounts to almost 3 tons of oil equivalent.

It is, of course, well known that one can use about 30 per cent less petroleum in a car of a given size by going to the high compression ratio diesel cycle instead of the spark ignition cycle. This also eliminates the lead in the exhaust. However, there are certain disadvantages of the diesel cycle, and as long as petroleum is cheap people will not be prepared to put up with them. In particular, the engine is considerably more expensive because it requires a high pressure injection pump for the fuel. A paper by Maltby *et al.** showed that the only effective ways of reducing transport energy in towns were: (i) to make it worthwhile for more people to use public transport, and (ii) to improve the fuel economy of the private car. In another paper at the same conference, J. D. Davis† discussed the possibility of storing the heat from diesel engines in a commercial vehicle fleet in a special heat storage unit, and then using this heat in externally heated engines. If three-quarters of the vehicles were diesel engines exhausting at some 700 °F, and if 70 per cent of the exhaust heat (which is 35 per cent of the fuel input) were captured and stored in lithium fluoride heat storage units, these heat storage units could then be used in the remaining one-quarter of the fleet to operate steam engines. He calculates that there could be a fuel saving of between 15 and 20 per cent, compared with a 100 per cent diesel fleet. The system would have the further advantage that the lithium fluoride heat storage units could be heated at the central garage with coal so that they could operate entirely independently of liquid fuel.

* Maltby *et al.*, 'An analysis of energy use in urban passenger transport', *Proceedings of the Newcastle Conference on Alternative Technology*, 1976.

† J. D. Davis, 'Prospects for low energy technology in road transport', *Proceedings of the Newcastle Conference on Alternative Technology*, 1976.

5.3 THE FUTURE OF ROAD TRANSPORT

5.3.1 *The future of the private car*

The number of cars per 1000 people has been rising steadily in all the countries of the world, with the USA and Canada at some 3–400, most of the developed European countries at 2–300, while the underdeveloped countries have a figure of the order of only 2–3 per 1000 people. It is quite certain that *there will never be a substitute for liquid hydrocarbons, pumped from wells, which will be nearly as cheap to obtain and as convenient to use, for road and air transport.* Not only does liquid hydrocarbon have a high calorific value and burn with some fourteen times its own weight of air, which it picks up as it goes along, but also it can be carried in a very light tank and pumped through very small pipes to the engine.

In the war we ran cars and lorries on solid fuel (anthracite, charcoal, or activated coke) converted to producer gas in portable gas producers. There was, however, an enormous sacrifice, in convenience, and weight, and cleaning the gas was very difficult. For private cars it was necessary to have a heavy trailer. It follows that when petroleum has run out it will be necessary to make liquid hydrocarbons from coal, or obtain them at great expense from shale or tar sands for road transport. Rail and sea transport can run on coal without much sacrifice once fully mechanized ways of handling the coal are developed.

Many people suggest that hydrogen could replace hydrocarbons as fuel for aircraft and for road transport. However, Table 5.2* shows that although hydrogen has a calorific value per unit weight nearly three times higher than liquid hydrocarbon, the problems of transporting it make it very much less satisfactory on a basis of equal energy output. Even for aircraft propulsion the reduction in weight is seriously offset by the enormous increase in volume of liquid hydrogen (nearly four-fold), so that it becomes necessary to equip the plane with two extra fuselages to carry the fuel.

In any case hydrogen is not a primary fuel but has to be made, inefficiently, at very high capital cost, from another energy source, usually electricity, which in turn requires high capital cost. *Thus the future of road and air transport is essentially one in which the cost of a suitable fuel will rise by a factor of at least two or three, compared with other energy costs, when the pumpable oil runs out.*

From this one can conclude that the most important work for the engineer, if the motor car is to be used in the future, is to obtain comparable

* K. Garrett, 'The future of the private car', *Chart. Mech. Engr.*, March 1975, p. 65.

TABLE 5.2 FUELS COMPARED ON A BASIS OF WEIGHT AND VOLUME

*Storage system**	*Weight of carrier and fuel (kg)*	*Volume of container (m^3)*
Hydrogen gas (at 13 789·5 kN/m^2	1027·00	1·87
Liquid hydrogen	161·00	0·29
Magnesium hydride (40 per cent voids)	316·00	0·31
Liquid ammonia	116·00	0·17
Petrol	68·40	0·08

* Figures computed for quantity of fuel equivalent to average tank capacity of American cars = 57·4 kg petrol or 20·6 kg hydrogen.

convenience with something of the order of one-third of the fuel consumption per passenger mile.

There are many other disadvantages of having more than one car to every five people; particularly, we have pollution, noise, built-in obsolescence, traffic jams, and accidents. It can also be a considerable strain on the driver in bad weather. On the other hand there is no doubt that for many purposes, such as family holidays or week-ends, the door to door carriage of luggage and other inconvenient objects, and because of its complete freedom from time-tables and shutdown at night, the family car is very satisfactory to human life. However, applying the principle that *only those things which can be available to all 8000 M people who will be in the world in the next century have a future,* one arrives at a very different picture of the role of the car in the future, from that in the USA at present. Certainly they will have to be light to save fuel, they will need to have very efficient engines, be completely non-polluting, and last for fifty years. There would probably be only one standard model and it is doubtful whether there could be more than one car to every ten or twenty people. Perhaps it will not be possible to have individual ownership, but a very flexible hire system, so that cars are used more fully round the week and not simply parked for 95 per cent of the time. Probably private vehicles will be banned from the middle of all big cities but the public transport system will be correspondingly improved to give greater mobility and convenience than can be obtained with a car at present.

I will now examine the problems of pollution and the future of road vehicle propulsion, including the bicycle, and the future of road transport for heavy goods.

5.3.2 *Vehicle air pollution and noise*

Vehicle pollution has been extensively studied because it impinges on everyone and because it has had severe consequences in the Los Angeles area, where there are very peculiar climatic conditions. The chemical pollution comes primarily from the engine exhaust; the other main possible source is the crankcase breather, but these are mainly hydrocarbon fumes and modern cars take these in through the air intake where they are burned in the engine. One way of reducing the impact of engine exhaust fumes, except in tunnels, would be to *raise the exhaust to the top of the vehicle* where there would be more air movement to disperse it. Most people who have studied the subject consider that the main sufferers are pedestrians and people in vehicles in traffic jams. Indeed, in a traffic jam in a tunnel it is necessary to turn off the engine because the ordinary spark engine makes a lot of carbon monoxide when idling, sufficient for people to use it when committing suicide. Although it is at a very low level, lead from the exhaust spreads over a very wide region and has been detected in polar ice.

The spark ignition engine is run with insufficient air, i.e., fuel rich, to give maximum power and good idling and acceleration. Pumping the accelerator squirts more fuel in through a special jet. During the over-run the fuel continues to pass in while the air is greatly reduced by the throttling action, so that high quantities of unburned petrol can be emitted from the exhaust, and it is possible to have an explosion in the silencer (so-called 'back-firing'). Another problem is that it is difficult to obtain an exactly equal mixture to the cylinders. For all these reasons combustion is incomplete except when cruising at part load and steady speed. The use of the choke makes the combustion particularly bad, and in an old engine combustion gases can escape to the crankcase. The unburned material consists of carbon monoxide, hydrocarbons (polycyclic hydrocarbons which are carcinogenic), and aldehydes. The use of low pressure fuel injection on a spark ignition engine can greatly improve the combustion and the fuel economy because it gives regulated fuel to each cylinder. However, high performance spark ignition engines will always be designed to run rich.

The diesel engine is considerably more efficient as the compression ratio can go up to 16:1 and the engine takes a full air charge at all throttle settings, being controlled by the amount of fuel injected, so that it is running lean under all conditions, except when the choke is used at starting or if the operator tries to obtain more than the designed maximum power by increasing the fuel input still further. The diesel has the further advantage that it does not require lead in the fuel, but it tends to be noisier, heavier, and more expensive, because the fuel has to be injected into the cylinders against the full compression pressure.

Lead has been introduced to the petrol of spark ignition engines at the rate of some 2 g/gal since the 1920s, and some 25–50 per cent of this lead is emitted in the exhaust as an aerosol. This is done to enable compression ratios as high as 10:1 to be used without 'knocking'. Some 2–4 mg/m^3 of lead in city air have been observed, and much of this lead will eventually get into the water and on to vegetables. Lead is, of course, a cumulative poison in the central nervous system, but it has never been conclusively proved that the measured levels have caused significant effects in people*. However, the engineer must take responsibility for such potentially bad consequences of a convenient method of improving performance.

In the USA there has been a great pressure to reduce oxides of nitrogen in the exhaust gases of vehicles. In this respect diesel and gasoline engines are about equal. This problem has arisen because in the Los Angeles area the combination of: (i) oxides of nitrogen, (ii) hydrocarbon emissions from refineries, (iii) the frequency of long lasting weather inversions that prevent mixing from the lower atmosphere, and (iv) strong sunshine, produces a photo-chemical smog which is extremely irritating to human eyes, causes crop damage, and the perishing of rubber. It is doubtful whether oxides of nitrogen are a serious problem in most other places, as the maximum concentration observed in a tunnel heavily loaded with cars is less than 0·1 ppm, whereas the industrial health limit for nitric oxide is 5 ppm† for an eight hour exposure. However, there is a possibility of long term effects from continuous exposure even at low levels. Oxides of nitrogen are formed almost entirely by the high temperature oxidation of nitrogen in vehicle conditions, although when coals or residual oils with high nitrogen content are burned in furnaces then the amount of oxides of nitrogen is considerably higher. In the country, oxides of nitrogen are probably deposited on the land by rainfall and help to increase the nitrogen required for protein formation in crops. It seems likely, therefore, that except in the Los Angeles area, this is not a serious problem.

* A team of researchers at the Boston Children's Hospital and Harvard Medical School have published a report in the *New England Journal of Medicine* (March 1979) giving the results of a correlation between lead in the milk teeth of children and their intelligence. Fifty-eight children had more than 20 ppm lead in dentine and 100 had less than 10 ppm. All confounding factors were eliminated and it was shown that the former performed significantly less well than the latter. (*New Scientist*, 5 April 1979, p. 5.)

† P. T. Sherwood and P. H. Bovers, *Air Pollution from Road Traffic—A review of the present position*, 1970, Road Research Laboratory Report LR 352. See also *The Automobile and Air Pollution*, Report to United States Department of Commerce, October 1967; 'Air pollution from road vehicles', Technical Committee of National Society of Clean Air, 1967.

On the other hand, sulphur in diesel fuel (0·3 per cent) can cause appreciable sulphur smell in the combustion gases as it forms sulphur dioxide or sulphur trioxide, leading eventually to sulphuric acid, or if combustion is incomplete, strongly smelling sulphur compounds. Petrol being a lighter cut of the petroleum fraction contains less than 0·1 per cent sulphur. Probably the most important contribution the engineer can make here is to reduce the amount of fuel consumed. Vehicles only contribute about 1 per cent of the total sulphur emitted by combustion, but the rest is all dispersed from high chimneys.

The other most evil consequence of the engineer's success in inventing the internal combustion engine is the question of noise. A great deal of research has been done to specify and measure traffic noise*. The unit known as the 10 per cent level (L_{10}) gives a fair correlation with dissatisfaction. L_{10} is the arithmetic mean of the hourly values, over a weekday period from 6 a.m. to midnight, of the levels in dB(a) just exceeded for 10 per cent of the time at 1 m from the façade of a dwelling. It is recommended that this should be less than 70 dB(a) for traditional buildings, that is, those which have not had special noise insulation. This level is reached if there are 6000 vehicles per eighteen-hour day on a road 30 m away, on which the average speed is 75 km/hr. If the average speed is only 50 km/hr this can go up to 40 000 vehicles per eighteen-hour day, but for higher average speeds it is impossible to achieve it. Noise screens can reduce the level by 10 dB(a) if they are placed close to the road and they are high, so that the noise is not diffracted round them. They have to be imperforate and have a mass of at least 10 kg/m^2. It is, of course, also possible to reduce the component of engine noise by massive shields, and damping round the engine, and the exhaust noise by better silencers. These tend, however, to have a penalty in overall mass and increased fuel consumption.

5.3.3 *Improved engines*

There is a very extensive literature on methods of improving the vehicle engine from the points of view of: (i) fuel consumption, (ii) enabling it to use all wider cut fuel, and (iii) reducing pollution and noise. These range from design changes of existing engines through to electric cars and steam engines. Modifications to the existing engine† include low pressure petrol injection to ensure exactly the same amount of fuel goes

* E. F. Stacey, 'Motorway noise and dwellings', *Building Research Station Digest*, 1971.

† See E. M. Estes, 'Alternative power plants for automotive purposes', *Proc. Inst. mech. Engrs.*, 1972, **186,** 125, and K. Garrett, 'The future of the private car', *Chart. mech. Engr.*, March 1975, p. 65.

to each cylinder, systems for vaporizing the petrol, high energy jet sparking plugs, and of course the changeover to the diesel engine.

Work has also been done on the stratified charge engine* where the fuel is burned initially in a fuel rich region and mixes with the rest of the air at a later stage, so that it is possible to obtain the performance of a fuel rich cycle but with the combustion efficiency of a lean cycle. This tends, however, to require high octane fuel, that is, a lot of lead. However, it has been shown by Professor M. S. Janota that the replacement of 15 per cent of gasoline by a mixture of methanol and higher alcohols (such as can readily be obtained from the conversion of natural gas) enables one to obtain the equivalent of 95 octane fuel (3 star) with no lead at all. Another line of experimental work has been the use of a camshaft with variable drive so that the time and duration of the valve opening can be altered under different performance conditions. The use of automatic gear shift in general tends to increase the fuel consumption since energy is wasted by the hydraulic slipping clutch.

Alternative types of engine include the gas turbine, the Stirling engine and the steam engine. The gas turbine* has been studied mainly because it can give a very compact engine of very high power and can run on a wide cut fuel and obtain very complete combustion. However, the fuel consumption is always high as it is not possible to have an adequate heat exchanger; also it is not possible to make very low power engines. It runs at very high speed and requires very accurately shaped turbine blades. It would appear likely, therefore, that it is a wrong direction for engineering to use something which was developed for high speed aircraft on the roads, even for very large lorries.

The Stirling engine† has been developed extensively and can give 40 per cent thermal efficiency provided there are heat exchangers both for the working fluid and for the combustion air. It will run on any fuel and can give very complete combustion. However, at the moment it is too bulky and too heavy to compete with the internal combustion engine. A large programme of research could probably overcome these defects.

Much work has also been done on the steam car. An intensive study is described by R. M. Palmer‡. The steam engine requires a condenser able to dissipate four to five times as much heat as a petrol engine of the same power. In this design study they rejected the possibility of using

* A. Curtis, 'Sharp turn for clean car thinking', *New Scientist*, 3 May 1973, p. 271. (See also Estes, *loc. cit.*)

† P. D. Dunn, 'The Stirling engine and its future', *Proc. Inst. Pat. and Inv.*, 1974, p. 11.

‡ R. M. Palmer, 'An exercise in steam car design', *Proc. Inst. mech. Engrs.*, 1969/70, **184,** Part 2A, No. 10, p. 195.

fluorocarbons because any leak would be so toxic. By using steam at 1500 lb/in^{2}abs at 900°F and condenser at 20 lb/in^{2}abs, they were able to obtain a Rankine cycle efficiency of 31·5 per cent. The engine had three cylinders with single acting pistons but an arrangement with tandem cross flow compound, that is each cylinder had two diameters and two pistons. There was a clutch and a two speed gearbox. It was calculated that at 20 miles/hr the steam engine would give 32 miles/gal, as against the petrol engine's 26 miles/gal, but at 40 miles/hr steam only gave 23 miles/gal against petrol's 29, and at 60 miles/hr steam was 17 miles/gal against petrol's 24. Thus, the steam engine gave the highest mileage, but at the lowest speed.

5.3.4 *Electric cars**

Many studies, theoretical and experimental, have been made on electric cars, and battery operated milk delivery vans are already in widespread use. The electric car has the fundamental advantages of silence, ease of starting and non-pollution. However, compared with the internal combustion engine the electric vehicle has the fundamental disadvantages that it must carry both reagents, and have a good conducting distribution system to take the current out of the poorly conducting paste, which stores the electricity chemically. Thus the lead–acid battery will always be enormously heavy compared with an internal combustion engine, and it is extremely difficult to get a range greater than 50 km on one charge. If the battery runs out away from the base the only thing to do is to get a tow from another vehicle. Moreover, not only is the first cost higher, but the expensive battery lasts only two to three years. Much work has been done on the sodium–sulphur cell, which has a much higher reaction energy per unit weight, and could overcome many of these defects, but not the one of running out of charge away from base. This cell operates at temperatures of 250–400 °C, but it would by no means be an insoluble problem to start it off by burning some fuel or to keep it heated all the time during discharging or charging. The lithium–chlorine cell is also being studied. Electric cars have been provided with regenerative braking, which increases the mileage before the battery is exhausted quite considerably.

A more likely way of operating electric cars in the future will be by means of the *fuel cell* in which a fuel is oxidized with air to produce

* *First Conference on Electric Vehicle Development*, 1977 (Peter Peregrinus, Stevenage), *The Economic Use of Electric Road Vehicles in a Changing Environment*, 1978 (Peter Peregrinus, Stevenage).

electricity. The only one that has been used commercially (mainly in space applications) is the Bacon cell, which runs at 200 °C and pressures up to 400 lb/in^2, but only works with pure hydrogen and pure oxygen. Experiments are being made on cells working with hydrocarbons (naphtha or natural gas) or alcohol and air, but these work at very high temperatures, breaking the fuel down to hydrogen and carbon dioxide by the water–gas reaction. United Technologies in the USA* are working on such a cell with a phosphoric acid electrolyte which will tolerate the carbon dioxide (unlike the Bacon cell which had an alkaline electrolyte). This is being studied as a static power generation system, but could lead eventually to a feasible electric car. Another very interesting possibility is the zinc–air slurry battery. The fuel is in the form of a slurry of metal powder carried in the electrolyte. It is oxidized with air, and when exhausted the cell can be drained and re-fuelled. The zinc can be reduced again at the filling stations.

5.3.5 *Hybrid vehicles*

One form of hybrid vehicle which is being used in towns in Germany is a trolley-bus which has sufficient lead–acid batteries to run for two to three miles away from the overhead wires, but when attached to the overhead wires can take enough current both to run itself and to recharge the batteries†. This is clearly a very attractive system for town and suburban public transport. The underfloor battery has a capacity of 230 A/hr at 360 V and weighs 15 per cent of the total vehicle weight. Regenerative braking is used. An electro-diesel hybrid is also being developed with no battery.

In April 1973, before the oil crisis, a conference was held at Queen Mary College, London, on the subject of fuel consumption and air pollution from vehicle engines‡. It became clear at the conference that the only practical method of producing a large simultaneous reduction in fuel consumption and pollution in vehicle engines, while preserving the convenience of the internal combustion engine, was the hydrid diesel-electric vehicle. The fundamental advantage of this is that one has a diesel engine, with just enough power for cruising, instead of having one of three times the cruising power in order to provide the extra power for acceleration; thus the engine is running at full power all the time the

* F. T. Bacon, private communication.

† E. O. Dietrich and K. Sahm, 'Operational data in Esslingen of the Duo-Bus', *Economic Use of Electric Road Vehicles*, Conference Publication No. 15, p. 65 (Peter Peregrinus, Stevenage).

‡ M. S. Janota (Ed.), *Vehicle Engines*, 1974 (Peter Peregrinus, Stevenage).

vehicle is moving. The extra power required for acceleration is taken from an electric motor operated by a battery which only needs to be three or four conventional batteries. In my Department at Queen Mary College, Dr Webb, Professor Janota, and Mr Bolandi are developing such a vehicle for a four-seater small car, with which it is hoped to get 100 miles/gal. The system has a conventional gearbox connected, by a conventional clutch, to a special 2 to 1 gear, the faster input to which is connected by a magnetic clutch to the motor, and the slower input to a diesel engine. In this way it will be possible to operate it in the modes: (i) diesel engine via gearbox to wheels, (ii) diesel engine plus motor to wheels, via the gearbox, and (iii) while stationary, diesel engine to motor, which then acts as a generator and recharges the batteries. The system will be provided with regenerative braking which will be efficient because of the use of the gearbox to speed up the motor in relation to the wheels. Studies are being made on the control system in an attempt to make it as easy to drive as a conventional motor car, with a small computer to tell it what to do under each condition.

Work is being done in America on diesel–flywheel hybrids, using an infinitely variable gear and a flywheel made of carbon fibres running in hydrogen to store the energy.

The possibility of operating buses with hydraulic transmission and regenerative braking using a compressed air cylinder to store the energy has been suggested. This could reduce the fuel consumption of the bus by 20–40 per cent because of the very frequent stops and starts.

5.3.6 *The bicycle*

Finally, we must consider the future of the bicycle. The bicycle is already a very sophisticated piece of engineering design, and very light weight racing bicycles with hard thin tyres provide a very efficient system for human transport. At the same time the bicycle gives valuable exercise which the user loses if he goes to work in a car. The principal problems are, firstly, the danger to the cyclist of riding in a mixed system with large lorries, buses, and fast cars, and, secondly, the effects of inclement weather. In a paper by Claxton* it is pointed out that the bicycle is quicker than the car in town for journeys up to 1½ miles, and the moped quicker for journeys up to four miles. A London traffic survey shows that the average distance travelled daily to work by a population of nearly 9 million is under four miles, and that the majority of those who travel a greater distance travel by train or underground railway, so that the

* 'The future of the bicycle in a modern society', *Jl R. Soc. Arts*, January 1968, **116,** 114.

average distance of those who travel by road is only 2½ miles. The maximum average car speed in London is only 10 miles/hr, while in the small town of Stevenage it was 20 miles/hr. It has also been pointed out that if 30 per cent of the people of town roads who use cars to commute to work, moved over to bicycles on a separate track, the contraction of the space occupied by the remaining cars would be nearly 50 per cent. Probably in the future the private motorist will not be allowed into towns at all, and the bicycle and public buses will be the sole means of transport. There is no doubt that the safety problem has to be seriously considered, as it has been in Holland where the flat country has encouraged the use of the bicycle. It may be possible, using new materials, to make a light weight bicycle or tricycle which is weather proof.

5.4 RAILWAYS

Compared with road transport, rail transport has several theoretical advantages.

(1) It is possible to operate it with very high capacity in terms of passengers per hour, or goods per hour.

(2) It does not require human steering and, therefore, can be operated much more safely in fog or slippery conditions and in moderate snow. It can also be totally automated if required, but in any case the passengers carried per driver can be very much higher.

(3) While there is a considerable noise from the wheels on the rails, the noise of a heavily loaded rail line is considerably less than that of a heavily loaded road, especially if the rail line does not have diesel engines but has electric drive. In any case rail lines have, in many countries, been installed for some time and are not situated close to existing housing in the way that new roads tend to be.

(4) *Safety.* In Britain nearly 8000 people are killed and over 3000 injured each year on the roads. Rail is very much much superior, according to Wickens*. In 1971 there was one fatality per 38·7 M traffic units on the roads and one fatality per 294 M traffic units on the railways. A traffic unit consists of passenger miles, plus freight ton miles. About 40–50 per cent of car occupant fatalities occur in car–truck collisions, so that one of the major factors is the problem

* A. H. Wickens, 'Transport has an elastic future', Institute of Fuel, Energy Brake or Break Conference.

of having large lorries and small cars mixed together on the same roads.

(5) Perhaps the most important of all is that trains can give considerably lower fuel consumption per passenger mile than any other

TABLE 5.3 (from Wickens*)

Vehicle	*System energy consumption (kJ/seat-km)*	*Average speed (km/h)*
Bus	170	60
Car	720	80
APT 250	600	190
Plane	2100	880

Consider the same comparison except that the space allocated to each passenger is the same for each vehicle, namely 1 m^2/seat. This comparison is shown in Table 5.4.

* *Loc. cit.*

TABLE 5.4

Vehicle	*System energy consumption (kJ/seat-km)*	*Average speed (km/h)*
Bus	550	60
Car	1030	80
APT 250	590	190
Plane	3500	880

The bus appeared to be efficient in the first comparison because the passengers were packed into a small volume. This could, if necessary, be done by other vehicles.

system, except a fully loaded long distance bus, as shown in Table 5.3, taken from A. H. Wickens' paper. Wickens points out that if one allows for the fact that the ground space of the train vehicle per passenger is much larger than that of the bus, and evaluates all the systems to the same space, namely 1 m^2/seat, the comparison becomes much closer, as Table 5.4. Allowing for the fact that the APT 250, the high speed advanced passenger train, has an average speed of more than three times that of the bus, then its fuel consumption becomes considerably better.

Thus, if trains are designed aerodynamically correctly, and without excessive weight or passenger space, they are the most

> efficient system, in terms of energy, for fast ground passenger transport. The railway train is inherently more energy efficient than a road vehicle because: (i) the air resistance per passenger is greatly reduced by the shaped nose and very slender body, and (ii) cast steel wheel on steel rails have much less rolling resistance than pneumatic tyres on a road.

Finally, one has the advantage that trains can operate on coal, either by electrifying them and burning the coal at a central power station, or by the use of coal directly in the engine. The original coal burning steam locomotives were very inefficient because the steam was not condensed and the expansion ratio could not be very great, even with multiple expansion cylinders. Studies at Queen Mary College indicate that we could make a modern coal-fired steam locomotive which was fully automatically fired, condensing, and combined the advantages of direct piston drive to the wheels with a low pressure turbine for the final expansion, to give an overall thermal efficiency as good as, or better than, a diesel-electric. Thus, one can predict that the railway system will play a greatly increased role in the twenty-first century, in both the developed and underdeveloped countries. The principal problem in laying new tracks is that the gradients must be very small.

Much work has been done on the magnetically levitated train, and on air bearings instead of wheels. These would greatly reduce noise and wear, but the levitation for air bearings would probably consume as much energy as the wheel bearing does at present. Research is going on into the possibility of using sodium–sulphur cells, so that an electrified train could run for some miles away from the part of the track which is electrified.

Existing rails are under used and, at present, in Britain, only the low revenue bulk transport goes on the railways, so that, with 18 per cent of the ton-mileage, only 6 per cent of the freight revenue goes to the railways. 200 M tons of general merchandise goes by road for distances greater than 80 km every year in Britain. When oil is really scarce all this long distance goods transport will go by rail. Schemes have been worked out whereby railways could carry goods on a two-tier system in which the 35 per cent top tier of goods would have overnight delivery on a nation-wide basis and the remaining 65 per cent would have delivery in a few days. The system could be operated with twenty major depots and 900 minor ones, with a simple loading and handling system at each minor depot based on a sideways gravity slope of 3 per cent so that the pallets or boxes would roll sideways from the higher loading platform on to the train and then from the train on to the lower loading platform. The delivery from the minor depot to the final destination would, of course, be by conventional lorry.

There is, of course, one goods transport system which uses very much less energy per ton mile than trains; this is the canal system. On these, a single horse could pull a load of 60 tons in a barge. The speed is naturally very low, but where canals have already been dug there is a considerable case for reviving them for the low revenue bulk merchandise such as coal, which is at present carried mainly by rail. Plans have been worked out for replacing the locks on the canal system by an air cushion tracked vehicle, which could be situated at the change of level and carry the barge up the hillside so it would float off in the canal at the higher level. Such air cushion tracked transporters could also lift barges out of canals and take them to factories situated 1 or 2 km from the canal.

5.5 SHIPS

Large goods-carrying and oil-carrying ships give a very low figure per ton mile if they are operated slowly, but the energy consumption goes up as the square of the speed. Some 5 per cent of the total oil of the world is used to propel ships at the present time, and some of the biggest ships use 200 tons/day. As oil begins to run out the first step will obviously be to use a coal-in-oil suspension, and, of course, it is possible to go back entirely to the use of coal, with pneumatic handling of grit-size coal. Experiments were carried out before the Second World War on ships running on pulverized coal, both carrying it as pre-ground coal in boilers, and with the mills on board.

No serious studies have been made on modifications to the conventional screw for propulsion because there has not been any great pressure for fuel economy. Possible lines for considerable fuel economy which are being considered at the moment are:

(1) The return to wind power with small auxiliary engines in case of need. This may be done by means of Flettner rotors, that is, vertical cylinders rotated by a small motor which act as sails, or by aerodynamically designed sails, made of modern materials, which can be electrically trimmed and furled to suit the wind and course required*.

* J. King, 'Appropriate technology in ships, wind powered ships', Newcastle Conference, 1976.

(2) Plate IV shows a model which has been studied of a new type of propeller which would reduce fuel consumption by some 30 per cent because it increases the mass of propulsive fluid by a factor of four or five. It would also eliminate cavitation and give improved braking.

5.6 AEROPLANES

In his paper 'The role of advancing technology in the future of air transport'* K. G. Wilkinson has pointed out that the whole of the spectacular development of the aircraft in this century has arisen because the advances were thought necessary for the maintenance of effective armed forces and the provision of efficient weapons. Since the Second World War it has been primarily for military purposes that we have developed jet propulsion, trans-sonic and supersonic aerodynamics, the structural use of titanium, radar for air traffic control, and inertial navigation systems. The steady increase in the speed of aircraft, up to a point well above the sound barrier has been, for the purposes of military 'keeping up with the Jones's', and each new increase in speed has been sold to the public regardless of whether it was the optimum for their needs. With the increase in speed there has been an increase in fuel per passenger mile, roughly as the square of the velocity, offset slightly by squeezing the passengers closer together, and by carrying greater number of passengers in an aircraft. Something like 50 per cent of the all-up weight of an aircraft taking off to fly the Atlantic is fuel, so that any reduction in fuel weight required would give a double advantage in that more pay load could be carried, as well as the direct saving. The determination to produce a supersonic passenger aircraft, as a spin-off from military aircraft, was quite certainly a movement above the optimum from the point of view of the travelling public, so that at best it is a luxury for people who have money to burn. Quite apart from the fuel consumption, there are several other reasons why the future of air transport lies in the subsonic region, perhaps even lower than the Mach 0·8, which is the normal subsonic speed for large jet aircraft at the present time. One of these arguments is that if one crosses the Atlantic in 2½ hours the 'jet lag', due to the time difference of about five hours, is imposed still more rapidly on the body, so that the gain in flight time is more than offset by the increased time the body needs to become used to the new time system. Other reasons relate to the sonic boom, which is an unnecessary increase in the ground noise of aircraft, already high enough as it is, and the fact that

* K. G. Wilkinson, 'The role of advancing technology in the future of air transport', *Jl R. Soc. Arts*, June 1977, p. 346.

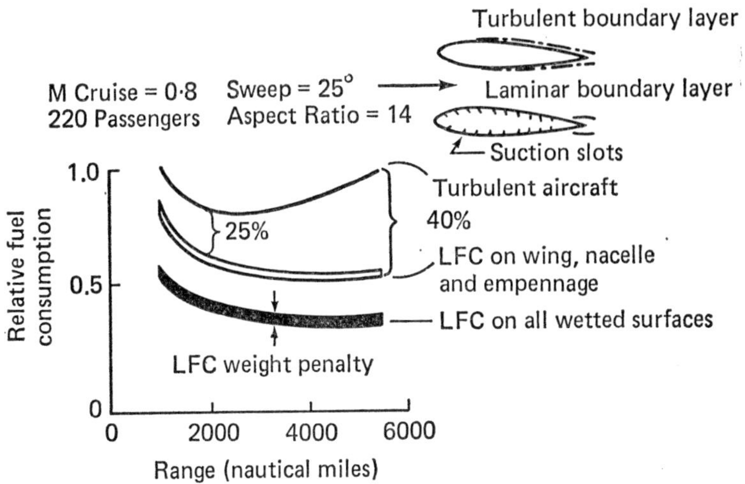

Fig. 5.4. *Laminar flow control fuel conservation potential*

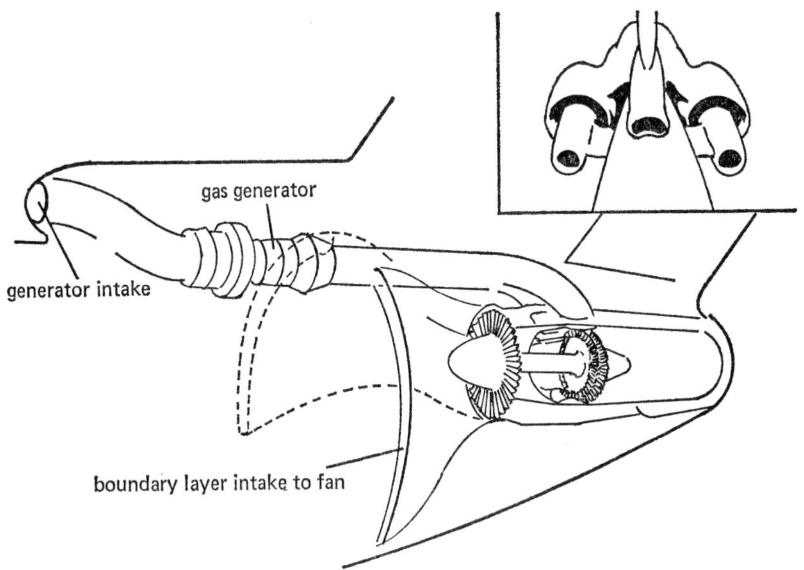

Fig. 5.5. *Boundary layer propulsion on fuselage*

aircraft designed for supersonic flight are not at all good from the point of view of the low speeds required for landing and taking off.

If the general philosophy of world futures outlined in Chapter 3 is achieved, then the principal future aircraft will be the 'air bus', carrying large numbers in large planes at relatively economic lower speeds, in order that all the people of the world can have some chance of overseas travel in their lifetime. On the other hand, there will be probably considerably less business travel than is the case at present among the developed countries. J. E. Allen* assumed 6 per cent annual growth in traffic up to the year 2005, so that seat-km rise from 8×10^{11} (1970) to 52×10^{11} in 2005. Such a growth may occur, but it is unlikely to be the result of an enormous increase in business travel in the developed countries; rather a much greater number of people all over the world who make one or two flights in a lifetime. Wilkinson also predicts a further four- to five-fold increase in world traffic in the next twenty-five years, accompanied by a doubling of average aircraft pay load, so that the actual number of movements of aircraft only doubles. Wilkinson considers composites can lead to a 25 per cent reduction in the structural weight of the aircraft, and he points out that experiments with suction slots on the surface of the wings and fuselage have been used to reduce the drag by maintaining a laminar boundary layer. In 1958 fuel was too cheap to make this worthwhile, but now that the cost of fuel is more closely related to its long term shortages, laminar boundary layer is being considered again (see Fig. 5.4).

The development of a geared, mid-fan, engine could give a 25 per cent fuel efficiency improvement by 1990, while the prop fan, with swept back blades, could give a 35 per cent improvement if it was accompanied by a 25 per cent lower cruise speed. This is likely to be the direction for the future as oil becomes more expensive. Some design studies (Fig. 5.5) have been made on boundary layer propulsion, in which the boundary layer air on the fuselage is taken in and accelerated by a turbine driven fan to provide extra propulsive fluid. It is believed that this might provide 6–7 per cent improvement in fuel efficiency. The ultimate line in this direction is to combine the wing boundary layer with the propulsive fluid flow (Fig. 5.6). In theory, this can improve the lift : drag ratio of the wing enormously, with a shaped wing and with the propulsive fluid sucked in along the lower edge of the wing and blown out along the upper side of the rear edge so that it provides both propulsion and circulation for lift†.

* J. E. Allen, 'Have energy, will travel', Royal Aeronautical Society, June 1977.
† M. W. Thring, contribution to discussion to 'A review of precious resources and their effect on air transport', Royal Aeronautical Society Spring Convention, May 1974.

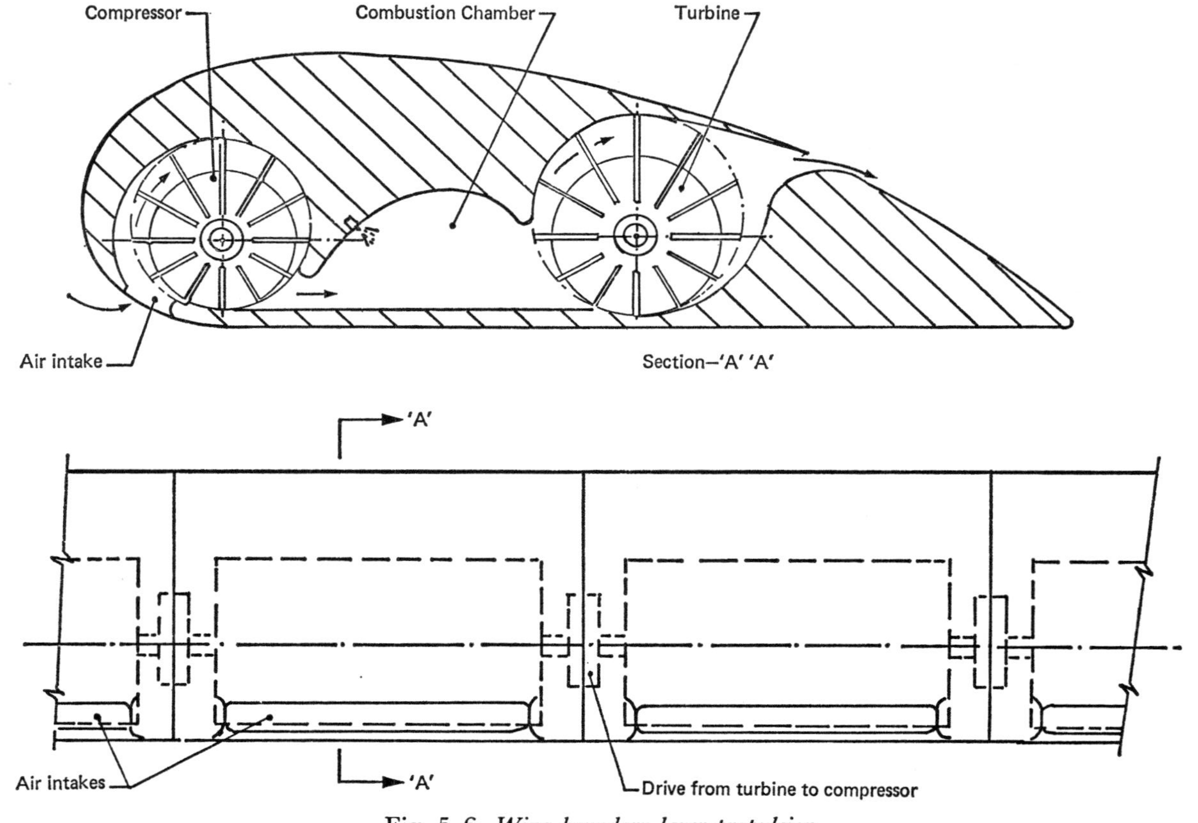

Fig. 5.6. *Wing boundary layer propulsion*

Proposals for totally different integration of air traffic with towns offer great potential for improving the most awkward part of air travel, getting to and from the aircraft on the ground, from one's point of departure. I have proposed a system in which the fuselage can be separated from the wings, fuel tanks, and propulsion system, and taken into the centre of the town on a monorail. After passenger interchange the fuselage is then taken to a central take-off terminal where a new wing system is fixed on and the aircraft is launched from a powered carrier on a monorail, so that the engines do not have to have such a big ratio of take-off power to cruise power. Such a system is, however, unlikely ever to be installed because it would require simultaneous changeover in so many existing cities.

Finally, one must mention the terrible noise problem to which people living near modern airports are subjected. E. J. Richards* has ascertained that each person flying into or out of Heathrow Airport is responsible for enough noise to subject 16·5 people to serious nuisance for an entire day. The number of people in fact seriously annoyed by Heathrow noise is of the order of half a million. There is evidence that aircraft noise can cause significant hearing loss and a significant defensive reaction which can be measured by muscular tension. There may even be an increase in the risk of heart disease.

5.7. ROLE OF COMMUNICATIONS

Communications are another area in which high technology is diverging further and further from the real needs of people, needs which will have to be satisfied in a stable world society. More and more sophisticated quadrophonic sound systems, television recorders, and so on, are being put on the luxury market, when most people in the world lack even an elementary education. Here again, much of the immense research effort is based on military 'requirements'. Money is relatively easily obtained for research that is connected with 'defence'.

There are three outstanding areas, where major changes will have to take place in the future, to which the engineer has a very large contribution to make.

(1) The problem of providing an adequate education, of all three brains, to 8000 M people as soon as possible within the limited resources of the earth.

* E. J. Richards, *New Scientist*, 25 July 1974, p. 189.

(2) The extent to which business and conference travel, which consumes a lot of energy and human time, can be replaced by better communication systems with visual contact as well as aural.

(3) The extensive use of timber for newsprint.

The first of these problems is one where the high technology of television, and even, possibly, satellite television, can play an increasing role. Of course, television can never replace direct human teaching, but it certainly can supplement it. The resources required for television communication to every village school in the world are not impossibly high in view of the great progress in micro-electronics.

The provision of the video telephone, at least for one conference room in every big organization, is again a case where high technology can improve human contact, although here again it is doubtful whether effective contact can be made unless the people have actually met at least once. Modulated light transmission in fibre optics for limited distances, or ultra short-wave radio transmission by direct beams, could achieve this, again without excessive use of raw materials or energy. This could reduce the amount of business travel very considerably. It is also likely that, as the tide of high technology recedes, the necessity of urgent business conferences on the other side of the world will be reduced, because, on a lower technology level, countries can be much more self supporting.

The excessive use of timber for newsprint will be greatly reduced once the luxury of advertising is no longer part of the economic system. Newsprint and other paper will certainly be re-cycled in the future, and methods of cleaning the ink off it, so that a white paper can be produced, are already being developed. The reduction in the amount of newsprint will probably make it no longer worthwhile burning domestic refuse as a heat source, since, at present, newspaper is the principal combustible constituent. It is doubtful whether it will ever be possible to make a microphotograph system, with a viewer, which is anything like as convenient for random access as a magazine or a newspaper, although if this problem could be solved the amount of paper used for these purposes could be reduced tremendously. However, there is little doubt that newspapers *can* be replaced totally in the future by a system in which any required page could be produced on a television screen. This would not allow one to read the newspaper in the train!

CHAPTER 6

Growing food and fuel for 8000 M people

6.1 WHAT ARE MAN'S REAL FOOD NEEDS?

Food is important in human life not only to provide the energy for activity, the materials from which our bodies are built, and the vitamins and trace elements necessary for health, but also because the daily process of eating meals can provide sensations which contribute as much as the enjoyment of flowers and trees to our emotional pleasure in life. The family, or friends, eating together has always been an essential part of right human relations. Thus, in addition to providing a good variety of food and a mixed diet containing a sufficient quantity of everything necessary, we have to provide good tastes and smells, food of genuine quality, and take as much account as possible of traditional ideas about texture, colour, and shape. Sir George Stapledon* goes so far as to suggest that the quality of food (its ability to enliven) is a criterion of good government. He also believes that 'supreme quality in all things is a gift from the centuries and not from science'.

However, science has given mankind the 'gift' of a vast explosion of population, so that to provide food of the right quality for all these people is clearly inconsistent with the use of the agricultural methods of the centuries. In this Chapter I shall study the steps which must be taken by the engineer (particularly the mechanical and biochemical engineer) to provide the necessary quantity, but I shall try, as far as possible, to take account also of the factors of quality, taste, and tradition. I shall base my figures for human requirements primarily on those quoted by N. W. Pirie†. An adult resting in bed consumes 70 W (i.e., 613 kWh/year, or $2{\cdot}21 \times 10^6$ kJ/year), people sitting consume 100 W‡, unenergetic walking 180 W, fast walking 380 W, and energetic sports up to 1 kW, briefly. Human muscles are about 25 per cent efficient, so we can produce work steadily at 95 W and peak briefly at 250 W. Well fed adults daily consume food of energy content 6 MJ (small sedentary women) to 17 MJ

* Sir George Stapledon, *Human Ecology*, 2nd edition, 1971 (Charles Knight, Tonbridge, Kent).

† N. W. Pirie, *Food Resources*, 2nd edition, 1976 (Penguin, London).

‡ 100 W average through the year is equal to 0·11 TCE (28 MJ/kg) per year.

(heavy labour). Pirie concludes that 6 MJ is the minimum daily average adult consumption for normal health and activity (1430 kcal/day in the units still used by dieticians), while the Ministry of Agriculture estimated an average requirement in Britain of 10·4 MJ (2475 kcal/day).

The energy in food comes from carbohydrates (starches, as in potatoes and bread, sugar, syrup, and treacle) and fats. In the rich countries the proportion of starches has gone down and that of sugar and fats gone up. Pirie remarks: 'With justification, the increasing use of sugar has been stigmatized as a drug addition'. Many doctors believe that the cholesterol in animal fats is a major cause of heart disease and advise patients with heart trouble to use vegetable fats instead. The world's highest incidence of bowel cancer is in the two countries that eat the highest proportion of meat, Argentina and Australia.

To feed 8000 M people will thus require

$$10 \times 365 \times 8 \times 10^9 \text{ MJ/year} = 3 \times 10^{13} \text{ MJ/year}$$

This could be produced by, for example, 1700 MT/year wheat, i.e., about 210 kg/person–year.

On the question of the minerals and vitamins in the processed food we eat in the developed countries, Pirie says: 'It can be regarded as a technological triumph that it is now possible to produce appetizing food that supplies the necessary amounts of the bulk components while being grossly deficient in some minor components'. Calcium and iron are deficient in cereal seeds, and milling and 'purifying' remove the small content in the outer layers. The use of wholemeal flour, or eating meat and liver, provides sufficient iron.

In general the application of existing knowledge could enable vitamins and trace elements to be available in sufficient quantities for all people. The problem is most difficult where people traditionally, or by shortage, live mainly on one foodstuff, such as rice. Since the quantities of vitamins needed are so small, it is feasible for them to be produced in factories in one country and exported to another on a permanent basis, unlike the main energy and protein constituents.

Pirie discusses carefully the question of how much protein is needed to build and maintain the body of a child or an adult, and concludes that the desirable protein level of children after weaning is more than 15 per cent, while protein deprivation of mother or young can lead to irreversible mental inadequacy. He concludes that it is now generally agreed that it will be more difficult to satisfy the world's need for protein than for any other dietary component.

He concludes that the protein consumption should be 1·5 g/day–kg of body weight. Thus a 70 kg person needs 38·3 kg of protein per year. From infancy to adolescence the requirement is 2–3 g/day–kg of body weight, and during pregnancy and lactation the mother needs an extra 6–15 g/day. This means that 38 kg/year can be taken as a desirable overall average figure per person, so 4000 M people need 150 M tons of protein per year and the 8000 M people likely to be alive in 2020 will need 300 M tons/year. Pirie says that at present the world's annual cereal crop contains about 100 M tons of protein, pulses and oil seeds 35 M tons, and potatoes and other roots 10 M tons, and there is about the same total amount in pastures and rough grazing. Thus, the total amount of protein grown on agricultural land is about 300 M tons/year. However, all the protein grown on the pastures and grazing land is fed to farm animals, as is much of the cereal and oil seed protein, and animals 'seldom return as much as a tenth of what they eat' as human food. One can conclude that if we develop processes whereby all the protein grown on agricultural land was directly available for human consumption, there would be enough for 8000 M people, but of course they would have no meat, eggs, cheese, or milk, i.e., they would all be vegans.

The situation is more complex than this because there are many different kinds of proteins which differ very much in nutritive value. The nitrogen in foods is mainly present in the twenty or more amino acids which form the proteins, and the protein content of the foods can be estimated as 6·25 times the nitrogen content when this is so. When the protein content of the food is low, however, and in leafy vegetables, much of the nitrogen may be in non-protein nitrogen form (NPN), such as urea. This is useless to humans, but ruminants can make protein from NPN with the help of the micro-organisms in the rumen.

Most proteins contain all the amino acids, but they vary enormously in their properties according to variations in the proportions of the amino acids and the way in which they are linked together. Eight of the amino acids cannot be synthesized in the human body, and so they are called the essential amino acids. The merits of different proteins as human food depends on the amount of these eight and their digestibility by enzymes in the human gut. Some proteins, such as hair, are totally indigestible, and others are made less digestible by food handling, e.g., drying or smoking fish.

Pirie says that it is not possible to divide proteins into good and bad; for example, formerly animal proteins were regarded as good and plant protein as bad, whereas it is now known that the protein in leaves and some tubers is as good as most animal proteins. Now it is recognized that different human contexts call for different proteins.

During growth amino acids are the building units of the body. In an adult the need for protein is to replace wear and tear, cope with the losses of protein resulting from the digestive processes, and provide for growth of finger nails and hair. Unlike carbohydrate, which is stored in the muscles, there is no store of available protein. Pirie concludes: 'to be sure of an adequate supply of all essential amino acids it is therefore necessary to eat more protein, as a rule half as much again, when most of the protein comes from less well-balanced sources'. Cereal and legume seeds (e.g., wheat, rice, and soya beans) are the most common ill-balanced sources. Wheat contains 10–12 per cent protein and, thus, a diet consisting solely of wheat would give enough protein if it provided enough energy, but rice contains less protein than wheat. Fats and sugar contain no protein.

In the case of cereal grains, the protein is concentrated towards the outer layer, so that a smaller grain contains a higher proportion of protein*. In apples the vitamin C is concentrated towards the skin, so again a smaller apple, and thus a lower yield/ha gives a better yield of nutrients and taste. Stapledon strongly advises against a rigid drawing up of diets which take no account of differences between individuals' needs and tastes, nor trains them to choose what is best for them. He is greatly in favour of diversification of production. It is clear that this should be ultimately the aim for all mankind, but that, as far as the less developed countries are concerned, the immediate aim must be *to grow an adequate supply of all the nutrients known to be necessary, on a permanent basis, within the earth's limited resources.*

6.2 THE USE OF FOSSIL FUEL ENERGY AND MINED CHEMICALS IN AGRICULTURE

6.2.1 *The present situation in the developed countries*

Table 6.1 (extracted from *Energy in Food Production* by G. Leach†) shows the fossil fuel input and the energy output of British agriculture and food processing in 1968. The units are 10^{15} J/year or 0·040 MTCE/year, and the population of Britain was then about 55 M. This table shows:

(1) The primary plant material grown on the farms contains some 2½ times as much energy as the fossil fuel used to grow the crops and to feed the animals on the farm.

* Sir George Stapledon, *loc. cit.*, p. 210.

† G. Leach, *Energy in Food Production*, 1975 (International Institute for Environment and Development, London).

(2) Mainly because animals are such inefficient converters of energy, and most of the crops are used to feed animals, the energy of the farm gate crops is about one-tenth of that grown; that of the plant foods actually eaten by humans is only just over half the farm gate figure.

(3) Animal fodder contains about five times the total fossil energy needed to grow it (so it uses sunshine very effectively), but the final animal food for humans has only about 7 per cent of the energy content of the fodder, i.e., the animals only convert one-fourteenth of the energy they eat for human use. They are, of course, more efficient than this in the conversion of protein. The overall result is that the energy in the fossil fuels used to produce human food, via animals, is about three times the energy in the food.

(4) Food processing, distribution, packaging, and selling use more than twice as much fossil fuel energy as is used on the farms.

(5) The total energy in all home grown food (which is 57 per cent of our requirements) is just over 10 per cent of the total fossil fuel energy consumed to produce all our food.

(6) The fossil fuel energy consumed in sea fishing is about twenty times the food energy of the fishes eaten. This figure rises rapidly as fish stocks are depleted and fishing vessels have to travel farther; for example, in the Adriatic from 1958 to 1971 fuel consumption per ton of fish landed increased by a factor of 2·6 (quoted by Leach from Levi and Gianetti, UN Food and Agriculture Studies Review, GFCM No. 53, 1973).

The advantage of eating plant products compared to eating animal ones is shown in the comparison of the fossil fuel energy requirements for margarine and butter in Table 6.2*. This table also shows the high energy requirements for distributing a product that must be kept cool, compared with sugar, which is a stable product.

In the USA and Canada the total energy usage for food production (34×10^9 J/year or 1·4 TCE/person–year) is about four times the total per capita energy use in Asia†, so that if all the world used as much per capita as North America, two-thirds of the total world energy use in

* O. Christensen, Food Technology Report, Fourth Scandinavian International Conference on Chemical Engineering, 1977.

† W. J. Chancellor and J. R. Goss, *Science*, 1976, **192,** 213.

TABLE 6.1 BRITISH AGRICULTURE; ENERGY INPUT AND OUTPUT (LEACH)*

Fossil fuel input to produce food for Britain				
To farms†	Primary plants	237		340
	Animals	103		
Imported fodder				53
Imported food				260
Fish‡				46
Processing	Fodder	51		527
	Food	476		
Food shops, etc.				139
Total all food				1300
Total food energy (57 per cent of requirement)				**138**
Energy flow to produce plant food				
Metabolizable energy from primary plant material				1116
Metabolizable energy of farm gate crop				119
Metabolizable energy of edible plant food				**65**
Energy flow to produce meat, eggs, etc.				
Fossil fuel input to produce fodder	British farms	103		207
	Imported	53		
	Processing	51		
Energy content of fodder	Home grown	951		1055
	Imported	104		
Animal output	Gross	94	Edible	**73**

* Units 10^{15} J/year (40 000 TCE).
† Breakdown of inputs to farms:

Buildings and services	52
Machinery	32
Fertilizers	82
Fuels and power	108

‡ Energy ratio for fish food production 1/20.

1970 would have been consumed purely for food production. In California, with a predominantly irrigated agriculture (using oil fired power stations), 22 per cent of the farm gate energy was used for fertilizer and 44 per cent for irrigation, whereas in Canada, for cereals and maize, fertilizer energy was 46 and 57 per cent, respectively, of the farm gate energy. *Five per cent of the world's natural gas production is used to fix nitrogen for fertilizer.*

TABLE 6.2 ENERGY CONSUMPTION (g oil*)

1 kg margarine	*g oil*		*1 kg butter*	*g oil*		*1 kg sugar*	*g oil*	
Cultivation of soya beans			*Cultivation and dairy farming*			*Cultivation of sugar beets*		
Cultivation in USA	110	206	Tractor	308	1484	Tractor	11	49
Transport USA–Denmark	96		Fertilizer	628		Fertilizer	28	
			Imported feedstuffs	185		Machine 'wear'	10	
			Dairy farming	252				
			Machine 'wear'	111				
Manufacturing			*Manufacturing*			*Manufacturing*		
Oil extraction factory	49	127	Dairy		268	Transport beets	10	219
Refinery	31					Sugar factory	209	
Margarine factory	40							
Skimmed milk	7							
Distribution			*Distribution*			*Distribution*		
Packing	47	169	Packing	47	169	Packing	9	26
Transport	8		Transport	8		Transport	8	
Retail shop	114		Retail shop	114		Retail shop	9	
Total		502	Total		1921	Total		294

* 1 g oil ≈ 10^4 cal or $4{\cdot}2 \times 10^4$ J; 1 kg oil ≈ 10^4 kcal or $4{\cdot}2 \times 10^4$ kJ.

One part of agriculture which is very strongly affected by fuel costs is glasshouse heating, which uses 25 per cent of the petroleum fuels used in British agriculture*. Better use of solar heating, better insulation, and the use of non-premium fuels, are therefore economically justified, even at the present fuel prices.

The use of dried grass or lucerne as a ruminant feed has the two advantages over hay that one obtains more than twice as much feed, especially protein, per hectare per annum, and that one is not dependent on good drying weather after the crop is cut. However, it does mean that one is using premium fuels (oil in Europe, natural gas in the USA) to evaporate water with an efficiency of less than 80 per cent. Since the grass when cut has only about 20 per cent dry matter, the fuel consumption is large.

Barley straw can be used direct as ruminant feed or treated with alkali

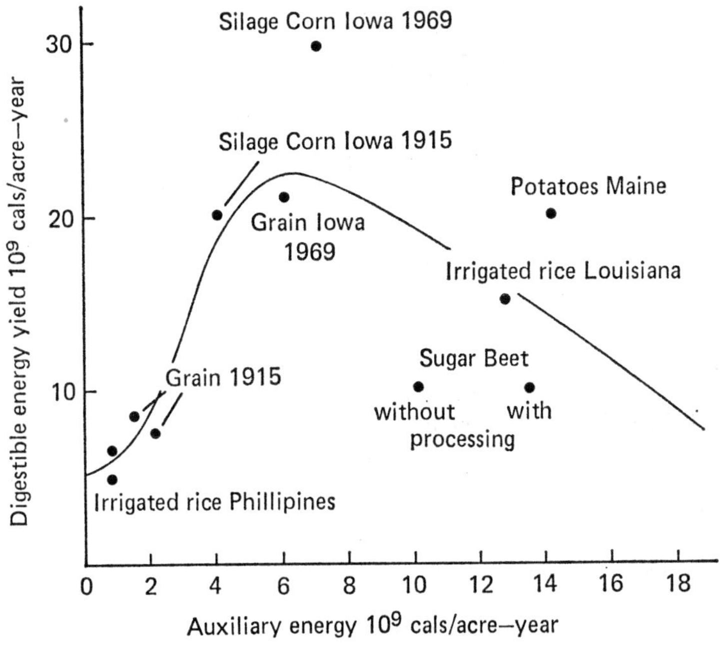

Fig. 6.1(a). *The result of using energy in agriculture*

* P. N. Wilson and T. Bigstocke, *Long Range Planning*, 1977, **10,** 64–70.

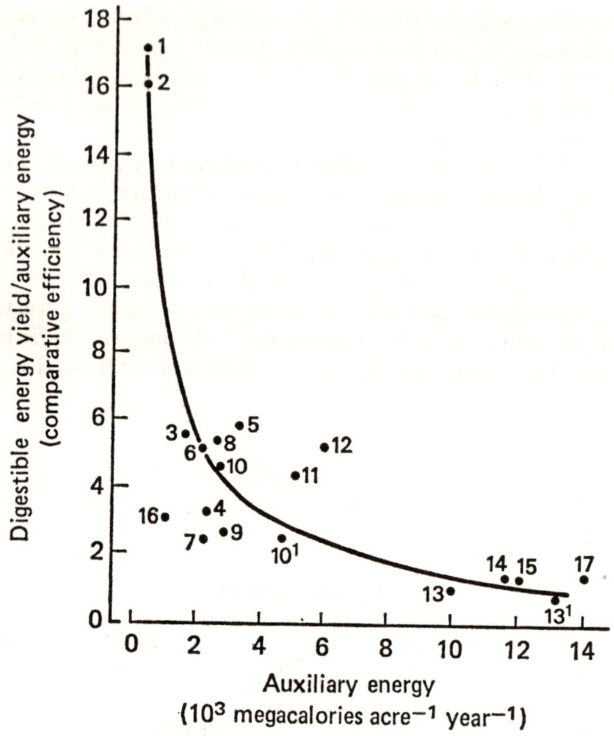

Fig. 6.1(b). *Recast of Fig. 6.1(a)* (*1 Mcal acre*$^{-1}$ = *10·3 MJ ha*$^{-1}$)

to break down the cellulose cell walls. Wheat straw is much less valuable, and rape straw is useless for this purpose, and they usually have to be burnt in the field as they are too tough to be ploughed in, as the stubble can be.

Figure 6.1 taken from 'Energy Requirements of Agriculture'* relates the fossil fuel energy use in agriculture to the digestible energy yield. The digestion may be by cattle, as in the case of the silage corn, in which case the resulting human food is only about one-tenth. This suggests that above about 6×10^9 cal/acre–year (25×10^9 J/acre–year or 1 TCE/acre–year) the digestible energy yield no longer rises, although one should only compare like crops, e.g., grain with grain and rice with rice.

Figure 6.2 shows how, over the years 1910–1969, as the use of fertilizers in the USA (measured as lb/acre of the nitrogen, phosphorus and

* Chapter 1 in M. Slessor, *Food, Agriculture and the Environment*, 1975 (Blackie, Glasgow).

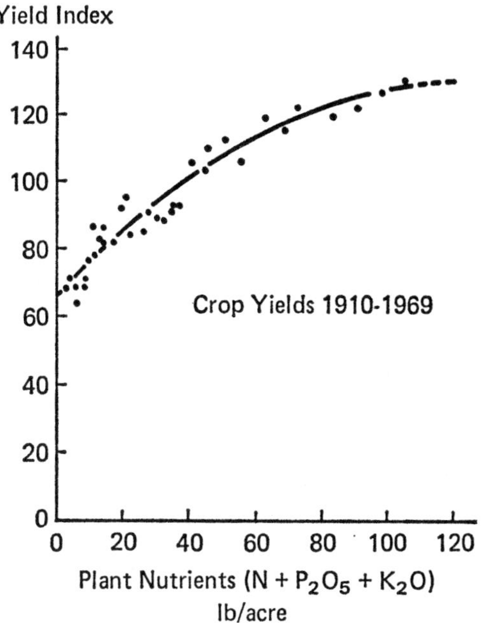

Fig. 6.2. *Signs of diminishing returns to increased use of fertilizers in the USA, 1910–1969*

potassium which are the bulk fertilizer elements) has gone from zero to 120, the yield has only doubled.

To complete this analysis of the present position, mention must be made of eutrophication—the excessive growth of undesirables, such as the blue-green algae resulting from chemical fertilizers being drained from the land by rainfall. This is especially bad in many of the inland lakes of the USA.

6.2.2 *The position of the less developed countries*

The Green Revolution was hailed, a few years ago, as the solution to the food needs of the less developed countries. These are the countries with the very high birthrates, largely due to the insecurity of their life. However, the Green Revolution has not had any widespread application. For example, by 1974 only 15 per cent of the rice land in South and South-East Asia had been planted with the new high yield varieties*. The high

* R. Allen, 'New strategy for the green revolution', *New Scientist*, 8 August 1974, p. 320.

yield varieties require large quantities of fertilizer and water, are much more susceptible to weeds and pests, and do not put their roots down so far. Very few local farmers can afford the technology of their use. The nitrogen fertilizer requires a lot of oil for its manufacture, and the rise in the oil prices has been a further nail in the coffin. It has been calculated that India would need 2½ new $100 M fertilizer plants every year, as one is needed to feed every additional 6 M people.

It is true that highly developed countries like the USA and Canada can, by using a great deal of fossil fuel and machinery investment, grow enough grain to supplement the whole world needs for one or two decades more. This, of course, can only lead to long term disaster when the world population doubles and the oil and natural gas become much scarcer. However, it is not even a temporary solution of the problem because: (i) the present economic system makes the less developed countries totally unable to pay for this grain owing largely to the low prices of the raw materials or bulk crops they export, and (ii) the richer countries make it uneconomic for their farmers to grow grain when they are in slump, and use the grain to feed their fatstock when they are in boom.

Everyone who has studied the matter carefully arrives at the following two conclusions*.

(1) The most immediate step is to find ways whereby the poorer regions of the world can grow enough food for an adequate diet for themselves. Since they have not enough resources to do the agricultural, engineering, biochemical, medical, and social research needed by themselves, the rich countries *must* devote enough resources to work in full cooperation with the poor countries, to find the solutions, and then contribute enough machinery to start the processes going on an adequate scale. It cannot be stressed too strongly that the solutions which are right for a rich country with abundant cheap oil, a temperate climate with regular rainfall, rich soil, and 150 years of industrial investment, are quite unsuitable for a poor country in the tropics. Each area of the world has its own problems and its own eating habits, but there is no doubt that if the rich countries were to devote a substantial fraction of the money they spend at present on weaponry and high technology (e.g., nuclear power, the space competition, and Concorde) to this problem, it could be solved before it is too late.

* For example: The Second Report of the Club of Rome; Colin Judge, *The Famine Business*, 1977 (Faber, London); G. Lean, *Rich World, Poor World*, 1978 (Allen and Unwin, London).

(2) In the twenty-first century, if we are to have a stable civilization, it will no longer be possible for the rich countries to have a diet in which meat and other animal products are the principal source of protein.

Both these conclusions illustrate Thring's Economic Principle 'whatever is right for our grandchildren is always uneconomic now and almost always impolitic'. It is uneconomic to help the poor countries to grow their own food and make their own machines instead of buying our machines for their cheap raw materials. It is impolitic to reduce expenditure on arms and high technology, and to ask people to eat less meat. Only in wartime are they prepared to make sacrifices.

6.2.3 *What can we do?*

For the overall problem of giving a good diet to 8000 M people in the next century the following smaller problems must be solved.

(1) We have to produce enough fertilizers on a permanent basis, especially phosphorus, potassium and nitrogen, to give good yields from all presently available agricultural land throughout the world, without using any fossil fuel except coal. Ultimately, even coal will be exhausted and we shall have to grow our own fuels.

(2) We have to provide enough suitable vegetable protein to be the main protein source.

(3) We have to provide enough machines and fuel to run them, so that full use can be made of the land, with 30–50 per cent of the population working on the land. The use of kitchen gardens and allotments may have to be greatly extended.

(4) We have to make full use of all products at present wasted, and recycle all necessary elements without pollution.

(5) We have to keep weeds, insects, and pests (e.g., rats) under good control.

(6) We have to find ways of controlling water to avoid the more disastrous effects of floods and droughts, and we have to reverse the process of desertification which is going on at the present time. We have to find ways of using swamps, inland lakes and hill land to the greatest extent.

(7) We have to consider where fish farming would make more human food available.

TABLE 6.3 PRODUCTION FROM 40 ha (100 ACRE) FARM IN EAST ANGLIA PAST, PRESENT, AND ESTIMATED FUTURE (WHEN OIL IS SCARCE)

	1875	*1975*	*2025*
Labour	3–4 men (full time)	2 part time men	2 full time men + 1 man (May to September)
Power	4 horses	1–2 tractors + electric motors (mains)	1 tractor + steam engine
Fossil fuel		16·5 tons oil (10 tons of this for fertilizer)	3 tons oil
Use of land			
Cows	12 ha		15 ha (grazing)
Ewes	12 ha		
Sows	4 ha		
Cereals	12 ha	40 ha	{7 wheat sold ha {8 stock feed ha
Lucerne			10 ha
Annual produce (sold ex-farm)			
Cows	(10 cows + 10 followers)		(25 cows + 10 beeves)
Beef	2 tons		3 tons
Veal	0·1 tons		
Milk	18 000 litres		90 000 litres
Cheese	0·14 tons		
Cereals	10 tons wheat	160 tons wheat and barley	28 tons wheat
Protein concentrate—(50 per cent protein) dry weight			8 tons
Fertilizer	All farm produced	N 4 tons P 2 tons K 2 tons	Fuel fixed N 0–1 ton Recycled from city refuse and night soil: N 1–2½ tons, P ¾–1½ tons, K ¾–1½ tons

Note: In the 2025 farm, cows graze on 15 ha in summer. In winter they eat the dried residue from lucerne after protein extract, and barley and kale from the 8 stock feed ha. The protein concentrate is used for human protein with suitable flavouring and treatment. Sheep may be kept in place of some cattle but will probably be on hill farms only.

(8) We have to reach equilibrium in sea fishing instead of making fishing more and more expensive by destroying the shoals.

(9) As oil runs short we shall have to grow more crops and make use of crop residues to produce substitutes for oil, plastics, wool, etc.

Turning to the special case of Britain (a developed country with a high degree of mechanization and a rich diet, growing only 57 per cent of its food) there is little doubt that our export trade in mechanized products must continue to dwindle and we shall have to become essentially self supporting in food. In particular, we shall not be able to import much meat, sugar, dairy produce, soya, or fish. We shall, thus, have to grow vegetables for most of our protein. We shall not be able to import any phosphorus for fertilizer, as the world will be seriously short.

These aims can be achieved without serious hardship, in spite of the increasing shortage of oil, if we do the following.

(1) Return to mixed farming (animal, cereal, vegetable) and crop rotation, as indicated in Table 6.3, which compares a farm such as my own as it was 100 years ago, is now, and should be in fifty years. Wheat straw could be used again for cattle bedding and fertilizer.

(2) Develop the Pirie leaf fractionation process to produce protein as a substantial source for human diet.

(3) Absorb a substantial fraction of our unemployed by a return of 30 per cent to regular work on the land. This is necessary to achieve good self sufficiency and full employment, but requires a Government subsidy equal to the unemployment pay.

(4) Develop the machinery for recycling. Also the use of waste materials, the use of city night soil, methane generation from cattle and pig dung, the use of carbon dioxide in plastic green-houses—all the processes that are at present uneconomic, but much less wasteful. Wind powered irrigation schemes, solar drying, better ground coverage, bringing marginal land into production, are all potentially valuable.

At present we have 12 M acres of arable land (of which 9 M acres are cereals, mainly wheat and barley) and 6 M acres of grass (rotated with arable), so we have about $\frac{1}{3}$ acre of cultivated land per person (55 M people)*. In addition there are 12 M acres of permanent grass and 17 M acres of uplands which could hardly be cultivated. We have a diet of

* K. Mellanby, *Can Britain Feed Itself?*, 1975 (Merlin Press, London), p. 10.

14·7 MJ/day (3500 kcal/day) average, and 75 g protein; while our basic requirement could be put at 11·3 MJ/day (2700 kcal/day) and 50 g vegetable protein.

Predictions of the future for British agriculture applying the present economic criteria are much less drastic, though broadly in the same direction. J. C. Bowman* says: 'the dairy sector, sugar beet, and oilseed rape are expected to expand, meat production to contract and the cereal acreage to remain largely unchanged'.

W. Williams† calculates that, with meatless diets, in Britain we should require 3 M acres to grow all our own wheat for bread, 1 M acres of barley for beer, 0·6 M acres of potatoes, 1·5 M acres of sugar beet, 0·7 M acres of vegetables and orchard fruit and 3 M acres of oil and protein crops. Total 9·8 M acres.

	M tons present	*No imports*	
		Milk and eggs	*All vegetable*
Protein in meat	0·35		
Protein in fish	0·1		
Protein in milk	0·5	0·8	0
Protein in eggs	0·07	0·14	0
Protein in protein-rich vegetables			1·0

Williams concludes that the milk and eggs alternative, probably with some extra vegetable protein, is desirable, and that we have plenty of land once we abandon the meat eating habit. Indeed we could afford to use land less intensively, back to half the present yield/acre, and thus eliminate the use of artificial nitrogen entirely (500 000 tons of nitrogen/year produced by 1·5 M tons of oil/year).

6.3 CROPS FOR FUEL, FABRICS, AND OTHER MATERIALS

In 1970 the annual human use of commercial energy was $2{\cdot}4 \times 10^{20}$ J ($5{\cdot}3 \times 10^{16}$ kcal)‡, whereas the annual solar energy incident on the earth's

* J. C. Bowman, 'An agricultural strategy for the United Kingdom', *Jl R. Soc. Arts*, 1977, **125,** 365.

† W. Williams, 'UK food production: resources and alternatives,' *New Scientist*, 8 December 1977, p. 626.

‡ W. J. Chancellor and J. R. Goss, 'Balancing energy and food production, 1975–2000', *Science*, 1976, **192,** 213.

surface was $3{\cdot}6 \times 10^{24}$ J ($8{\cdot}8 \times 10^{20}$ kcal) and global plant growth converted $2{\cdot}4 \times 10^{21}$ J. Thus, if we could utilize one-tenth of the present conversion we could supply mankind with as much energy as was used in the world in 1970, i.e., 10 000 MTCE/year. For 8000 M people this would give 1·25 TCE/person–year. This is less than the 2 TCE/person–year suggested in Chapter 4, and we have to conclude that *we shall require to grow more fuel crops in many heavily populated areas of the world* and probably irrigate all the desert and semi-desert areas partly to grow fuel crops for export. This possibility will be considered in section 6.8. We shall certainly have *to develop ways of increasing the efficiency of solar conversion on large areas of the earth that are already covered with vegetation.* There are many ways in which fuels can be obtained from agricultural products.

(1) *Direct combustion*, preferably preceded by a natural drying process. Straw, bagasse (sugar cane residue), wood, even dried animal dung, have been burnt. All the potassium and most of the phosphorus are retained in the ash, but the little nitrogen in this part is lost. Briquetting, baling, or cutting into short pieces is often necessary before these fuels can be conveniently used. Wood has, of course, been used as a combustible since fire was discovered. The old English system of 'coppicing', in which a tree stump, called a stool, retained its roots and sent up half a dozen poles which would be cut every seven years or so, is a very efficient method of converting sunshine.

(2) *Methanol* was originally called 'wood alcohol' because it was produced by the low temperature pyrolysis of wood as a by-product of charcoal formation. Charcoal is, of course, an excellent fuel, but it has many more valuable uses, especially because of its enormous internal surface area.

(3) *Bacterial conversion* such as fermentation. This includes making ethanol from sugar cane, cattle feed yeast protein from straw and other forms of cellulose and lignin, a human food yeast protein from coconut milk, and methane from pig slurry by anaerobic fermentation*. Ethanol from cane sugar can be used directly as fuel in modified spark ignition engines, or it can be dehydrated to

* The anaerobic fermentation of animal wastes to produce methane removes hardly any of the nutrients from the manure, so the nitrogen, phosphorus and potassium can all be returned to the land. 187 M tons/year of animal wastes are currently produced in the UK. The slurry would have to be used in very local plant which has a high capital cost and requires heating to 35–37°C (P. N. Wilson and T. Bigstocke, *loc. cit.*). In Mexico solar heating is being used to produce this temperature.

give ethylene which is the starting point for the manufacture of various chemicals and polymers. J. A. Lane (private communication) of New South Wales, who has done experiments on the matter, estimates that in Australia a 2 hp engine could be fuelled and lubricated (castor oil) on the products from 0·2 acre, whereas each horse would need 1 acre of good pasture to keep it 'fuelled'. This is because one only has to feed the engine when it is working!

(4) While it is possible to produce oil by the hydrogenation (reaction at very high pressure and moderate temperature with hydrogen in the presence of a catalyst) of coal or tar, many kinds of agricultural refuse make a much more reactive feedstock. Cotton plant residues were used successfully in studies in my laboratory in Sheffield.

Brazil is aiming to produce 20 per cent of the country's total fuel consumption by ethanol from sugar cane and cassava (20 M m^3), and, by using a heat exchanger, this can be burnt in their cars with no loss of efficiency. Large areas of cheap land are available for machine cultivation, mainly by cutting down the forests of the Amazon.

M. Calvin* is studying various latex producing shrubs (the rubber tree is the best-known one)—they store the sun's energy directly as hydrocarbons—to find the 'gasoline tree', which will grow in very arid regions. The plants contain between 2 and 10 per cent of the total fresh weight as hydrocarbons, so a yield of more than 2 bbl of oil/acre–year could be obtained.

J. A. Lane has also told me of the fibre called *Vicara* which was manufactured before 1957 from zein protein extracted with alcohol from maize grain. This is the least nutritious protein (new varieties of maize which are much easier to grind and do not cause pellagra are being developed). He calculates that the yield/acre of this vegetable protein wool is five to twenty times that of sheep grazing pastures. It has a high moisture absorption and a warm soft feel like wool and resists rotting by bacterial and fungal action much better than cotton and rayon.

6.4 THE DEVELOPMENT OF NEW CROPS

The sole source of energy for plants is sunlight through photosynthesis, but they use it very inefficiently. As we saw in section 6.3, the total annual solar incident energy is 1500 times the land plant growth, so,

* M. Calvin, *Sunworld*, No. 5, August 1977, p. 2.

allowing for the fact that about four-fifths of the earth is water, 1/300 of the land sunlight is converted into plant energy. Much of the sunshine on the sea is also converted into plant energy so that, in fact, most of the carbon dioxide produced by combustion and animals is converted back into oxygen and released over the oceans.

One-quarter of the energy in sunlight is reflected because the leaves are not black, and only half of what is absorbed is of suitable wavelength. The lost energy evaporates water from the leaves (thermal transpiration) which is valuable, up to a point, because it brings nutrients from the roots to the leaves, but in very hot climates the sunshine is so intense that there is an excessive loss of water just to keep the leaves cool. This energy also causes convection which brings carbon dioxide to the leaves. Pirie* calculates that on a sunny day in Britain 2 MJ/hr–m^2 of useful solar energy is available, which, with 4 per cent efficiency, could produce 5 g of dry matter (calorific value 17 kJ/g), removing 7·3 g of carbon dioxide from the air and requiring all the carbon dioxide in 12·4 m^3 of air (carbon dioxide in air 0·59 g/m^3). Since plants can only use a small fraction, it requires several 100 m^3/day of air to penetrate each m^2 of crop. However, it is only on still days at times of vigorous growth combined with full ground cover by leaves, or in greenhouses, that carbon dioxide starvation reduces growth rate. In actual fact the cereals and sugar beet with spikey upright leaves can reach 4–5 per cent efficiency *when they have established complete ground cover*, but they are in this condition for only a small fraction of the year. Many weeds can photosynthesize in the early spring sunshine in Britain because their roots are frost hardy, and if we could use this property better use of the annual sunshine could be made. The development of short strawed cereals has already put a bigger fraction of the sunshine into the grain—the straw weighs about the same as the grain now.

The crops grown in the temperate climates have what is called the C_3 system of carbon assimilation in which part of the fixed carbon is breathed, out again as carbon dioxide, whereas many tropical species have the more efficient C_4 system. Work is being done to try to develop C_4 crops which will grow well in temperate climates. Our season is rather too short for maize varieties used at present. There are many other ways whereby plant breeders can improve British output†, such as

(1) searching the world for species that might have value if cultivated in our climate,

* N. W. Pirie, *Food Resources*, 2nd edition, 1976 (Penguin, London), p. 131.
† J. P. Hudson, 'Food crops for the future', *Jl R. Soc. Arts*, 1976, **125,** 572.

(2) developing nitrogen-fixing root nodules on non-leguminous crops like wheat, or enabling these crops to obtain nitrogen from free living nitrogen-fixing bacteria,

(3) developing perennial wheats using the same roots year after year,

(4) developing apple trees which are planted 0·3 m apart and kept to a single stem which crops heavily in the second year and is then mown off leaving the tree stumps only. This gives full cover of leaves from the outset.

In the tropics with no winter frost problem, crops can be grown all the year round and several successive crops can be grown in the year. The problems are mainly water, excessive heat, lack of fertilizer, and lack of knowledge. Intercropping, such as maize with beans, maize with peanuts, and cabbage with tomatoes, can benefit the primary crop by reducing pest or weed attack. The International Rice Research Institute is now relating its work as much to cooperation with the local farmers as to the production of higher yielding varieties.

Pirie considers that algae are unlikely to become a widespread human food because they require the carbon dioxide to come via the water in which they grow, but that algae grown in sewage in conjunction with bacteria can help the purification of the sewage and the mixture can then be curdled off, filtered, and used as animal food. Medieval monks ate carp that lived on sewage.

The production of food from materials at present wasted will clearly have to be done regularly in the future. Among the many wastes from which we can make protein-rich human foods are:

(1) whey from cheese—this can produce excellent proteins for baby foods*,

(2) coconut milk†; 6000 tons of yeast protein could be produced annually by fermentation in the Philippines,

(3) residues from oil seeds; Pirie says that the oil seeds from protein-deficient countries supply much of the protein fed to pigs and chickens in rich countries like Britain.

6.5 NEW MACHINERY

The energy used in ploughing is very substantial and direct drilling processes, in which the weeds are killed with spray and no ploughing is

* O. Christensen, Food Technology Report, Fourth Scandinavian International Conference on Chemical Engineering, 1977.

† N. W. Pirie, *loc. cit.*, p. 178.

done, are being used in both the developed countries (e.g., for rape) and in the underdeveloped ones. The International Institute of Tropical Agriculture (Ibadan, Nigeria) has developed a battery powered micro sprayer so that a farmer can kill the weeds with 2 kg/ha of herbicides applied from low down. R. Wijewardene of that Institute is trying to develop a three-wheeled vehicle of 3–5 hp which is primarily a transporter but can also spray 6 m, plant two rows and, with the operator walking, can do 1 m rotary tillage and 0·9 m rotary weed slashing. As liquid fuel is very expensive in these countries it could be battery propelled, run on ethanol, or have a steam engine burning local biomass.

The ploughing tractor has the disadvantages that it pans the soil and has considerable energy losses in wheel slip. It is necessarily very expensive and requires large wheels with expensive tyres. The 'Snail'* pulls a plough (formerly ox drawn) by winching a rope from a self propelled winch which has a sprag that is driven into the ground by the pull. It can give a pull of 5·5 kN (100 lbf) and has a 6 hp diesel engine, but, as the name implies, it is slow (0·56 m/s), requires two men, and is only ploughing for about half the working time, so it requires 8·5 man-days/ha.

My own proposal is to use a walking machine (see Plate V) which can have different feet according to the soil. The Snail was developed for countries where it is desirable to cultivate the soil at the end of the dry season when the ground is very hard, to make full use of the rains. Any system which requires the winding and unwinding of a cable is always slower and requires at least twice as much manpower as a riding tractor pulling one or more ploughshares. The increased manpower may not be a disadvantage in the future or in less developed countries, but speed of working is always important when the climate is the least bit unreliable or the exact date of the monsoon cannot be predicted. The old system of a pair of steam engines winching a plough backwards and forwards across a field shared with the horse the advantage of not 'panning the soil' as a heavy tractor does. The heavy tractor can also go in up to the axles in wet clay because wheels are continually changing the point of contact and so never really establish a static grip. Legs with feet and caterpillar tractors, on the other hand, have a load-bearing contact area, or foot, which is completely stationary while the body moves forward over it. The foot, moreover, does not rely on tyre tread to give it grip because the 'heel' can penetrate to a considerable depth before it is picked up. If a mechanical walking machine can be developed, which is

* C. P. Crossley and J. Kilgour, 'The Snail: a new cultivation system for small scale farming', *Proceedings of the Newcastle Conference on Alternative Technology*, 1976.

cheaper and simpler than a caterpillar tractor, it will be of immense benefit to farmers all over the world and save much fuel in the high traction operations. My proposal shown in Plate V relies on conventional drive chains for the whole movement.

Barley straw can be nutritionally improved by treating with alkali to break the cell walls, and it then becomes a valuable constituent of cattle feed. The machinery for this process has been installed in the same factory in Yorkshire, by British Oil and Cake Manufacturers—Silcox, which can operate the leaf fractionation process on lucerne in the summer. This leaf fractionation process has been pioneered over the last thirty years by N. W. Pirie, primarily to provide good protein for humans. Any young fresh green leaves can be used, including even fresh water weeds, and the quality of the protein is as good as animal protein, and better than seed proteins such as the soya bean. The yield/ha is of course many times as great as that of animal protein, and hence it can be produced very much more cheaply. Up to 6 tons of crude protein could be obtained per hectare per annum in the wet tropics and 2 tons in a climate like Britain. At present the main obstacles are the taste of the protein and the food habits of the people who need it. Screw presses are used as the 'mechanical cow' to extract the protein-rich juices from the leaves, but no one has developed a satisfactory small, single-screw press which could be used with

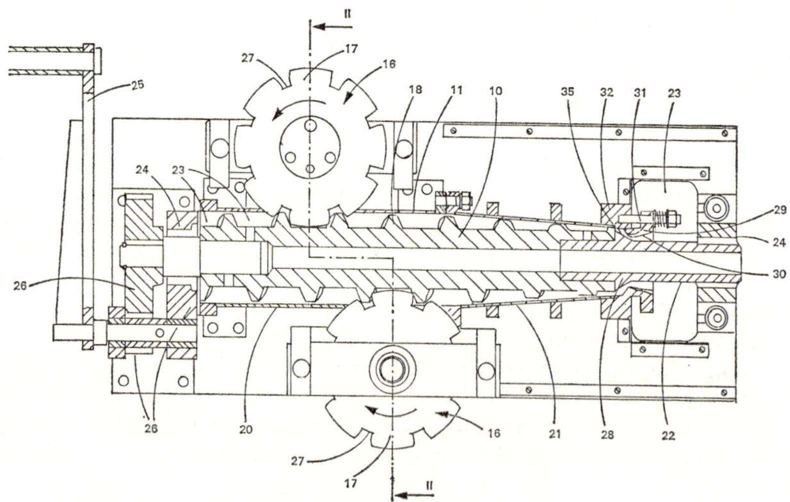

Fig. 6.3. *Single screw press*

low power operation in villages, and which would satisfy the conditions laid down by Pirie of rupturing the cell walls and squeezing the juice through a very thin layer of fibre to avoid filtering the protein on to cell walls. A design of the author's for a single screw press with a pair of worm wheels to act as a nut and prevent rotation of the material in the screw thread is shown in Fig. 6.3.

Legumes like lucerne have a root nodule system that binds its own nitrogen and so can produce protein with less or even no addition of artificially fixed nitrogen. The BOCM-S process, with which the author was associated, uses lucerne for this reason and collects it from farms up to seven miles away, cropping four or five times during the five or six summer months. Roughly half the protein (protein is about 20 per cent of the dry matter) is extracted in the juices in a twin screw press, curdled by heating nearly to boiling point, separated in a continuous belt filter, and dried. The residual fibre is still a good ruminant feed and only requires the evaporation of one half as much moisture. In the USA the protein is curdled in two stages, at 60 °C and 80 °C, and the second curdling contains about half the extracted protein and is white and sufficiently free from the strong flavour of the lucerne, to be usable as a protein additive for human food.

The process of multiple evaporation was originally developed to save steam in the concentrating of sugar; similarly the process of multi-flash distillation was developed to produce distilled water for drinking. In both processes the latent heat of evaporation is used several times over and most of the water condenses before leaving the system. In the process of vapour compression drying, all the evaporated water is condensed and the energy to run the process is provided by the engine that compresses the vapour to raise its temperature, so that the latent heat can all be recycled. This energy may be as low as one-sixth of the latent heat and this process can be used for drying solids. The possibility of using it for drying grass and other agricultural products, where a lot of moisture has to be removed, and for drying sewage sludge after water has been pressed from it, will all become worth serious consideration as the price of premium fuel goes up.

Solar energy has been used experimentally for grain drying by drawing air through a layer of grain exposed to the sun. In Britain, we have one-third as much solar energy as in the tropics, but it could be worth using it for some drying purposes if a solar collector of sufficient area could be built at a low cost, and provided some auxiliary heating was available when there was no sun. In any case a large fan would be needed and the waste heat from the engine that drives this could be used at all times.

In the developed countries the new machines will primarily be concerned with *producing more human food than at present with very much less oil and other fossil fuels* and with *establishing a more permanent equilibrium relationship between man and the environment.* However, much more manpower will be economically available, both on a permanent basis, and on a temporary seasonal basis, for harvesting (see Chapter 7). Developments, such as the combine harvester, which make us much less dependent on the vagaries of the weather will continue to be essential.

In the underdeveloped countries it is in general necessary to study the particular needs of each area (as is being done by the Intermediate Technology Development Group) so that machinery can be developed whereby each area can become fully able to grow all its own food and fuel.

6.6 FERTILIZERS AND PESTICIDES

The production of artificial nitrogen is the biggest fuel user in the agricultural industry, and it is probable that as fossil fuel costs go up this will become too expensive. We shall have to recycle all nitrogen by composting all animal and vegetable waste material, including dried blood, and eventually by developing a process whereby the soluble nitrogen in urea can be recycled. We may have to develop hygienic lavatories using much less water. The return to mixed animal/cereal/vegetable farming and crop rotation will help, and so will the development of nitrogen fixing bacteria for crops other than legumes.

As far as the other principal fertilizer elements are concerned, phosphorus and potassium, we shall have to reduce the amounts washed off into the rivers and develop methods of extracting them from water and even from the sea. The use of dried sewage sludge (from which the heavy metals have been extracted for recycling) on the land will also have to become universal. Sewage sludge could provide a protein source for poultry, ruminants and non-ruminants*. A continuous high-pressure filter press has been developed to bring the water content of flocculated and spin dried slurry down to a reasonable figure for final evaporative drying by Edwards and Jones of Stoke-on-Trent.

Weed killers developed from petro-chemicals are of enormous value to the modern cereal farmer, and even such persistent weeds as couch grass and wild oats can be controlled. These do not use very much oil and will probably continue to be necessary but, of course, are less essential

* *The Chemical Engineer*, May 1977, p. 342.

where mixed farming, crop rotation, and hand hoeing can be economically justified.

The petro-chemical insect killers are very effective and, again, essential to the farmer who grows a single high-yield crop on a large field, but there are severe problems with danger to human life from spray drift, especially aerial spray. The use of very low hand held sprays in the tropics, developed by the International Institute of Tropical Agriculture, can use the spray material much more economically and avoid the drift danger. Promising experiments are being done on the biological control of insects, by interfering with mating habits or introducing killer insects.

6.7 WATER CONTROL

Every year one reads of thousands of people being killed by floods and, in other areas, by famines caused by drought. These are problems which must concern deeply the engineer with a conscience, and there is no doubt that if a small fraction of the engineering ability and resources devoted to weaponry was diverted to these problems, they could be solved.

Water storage by dams has been used extensively and is valuable in providing storage from a rainy season through the rest of the year, or even from a wet year to a dry one. But it can alter the agricultural situation harmfully by preventing the carrying down of rich silt, and there is a real danger that this silt will fill the dam in 50–100 years, so that the storage capacity of the dam will disappear. Thus, while a large dam in the mountains may be the cheapest way of storing water when built with big machinery, and certainly makes hydro-electric power available, it may not be the best way of averting drought problems in all disaster areas. The possibility of small reservoirs all along the river's length, protected suitably to reduce evaporation, or even underground (as in North Africa in Roman times), should be examined in certain cases. In many places, e.g., Nigeria, the monsoon rain is available a few metres underground in a porous land mass, and the problem is to pump sufficient quantities to the surface for irrigation. Experiments with a Savery type solar pump have not shown adequate efficiency, and the use of a solar steam engine to drive a pump is now being worked on.

Experiments near Eilat in 1949* have indicated that *many plants can be grown on barren sandy hills irrigated with highly saline water* (2·3–6 g/litre). In the midst of salt-tolerant plant associations, non salt-tolerant plants

* H. Boyko, 'Sand-deserts and granaries of the future', *Futures*, March 1969, p. 191.

could be grown, provided the soil was sand or gravel so that there was good percolation and irrigation. Food and fodder plants, fifty-five tree species, and many industrially useful shrubs were all grown. Some cereals and fodder grasses could be grown irrigated with sea water up to 40 g/litre total dilution of solids. A wheat has been found in India which thrives well under an irrigation of 20 g/litre total dissolved solids, of which 15 g/litre was sodium chloride. The possibility of pilot farms based on these results should be turned into reality.

C. M. McKell of Utah State University has proposed* the utilization of the shrubs that grow naturally in the arid lands for industrial raw materials, such as latex, and for sheep and other animal grazing, instead of their destruction.

It is, however, clear that, ultimately, we have to find a method of restoring the desert areas of the world to natural growth. I say restoring deliberately, because many areas that are now desert, such as parts of the Negev, Gobi, and Sahara deserts, were fertile and used for agricultural purposes thousands of years ago. Towns flourished around ancient methods of collecting dew, reducing evaporation, and even underground storage, but wars and the depredations of goats have destroyed the vegetation over vast areas. It is even possible to change the whole climate of a desert area once trees and ground cover have been re-established. Experiments in very hot areas of Africa† have shown that two-thirds of the water needed by a plant is used for thermal transpiration through the leaves to keep it cool in the hot sunshine, and that the water needs of crops can be halved by shading them from the full sunshine using white, fine mesh nets that reflect most of the sunshine but allow the air to circulate freely. This is a development of tree-shading and cactus-shading methods which can be extended very widely.

The only method of producing fresh water in large quantities to irrigate deserts which does not require fossil fuel to operate it is the solar distillation of sea water. The use of waste heat from power stations and multiple effect evaporation are feasible to supply drinking water to industrialized areas, but are not feasible for agricultural purposes because of the fuel costs as well as the capital cost. Solar stills with glass roofs have been used for over 100 years, and much recent research has been done on these, and on plastic roofed ones, in Australia and the USA. So far, however, the capital cost and the difficulty of installing in desert areas have been too high for agricultural development, but plans are being made for a pilot

* C. M. McKell, 'Shrubs—a neglected resource of arid lands', *Science,* 1975, **187,** 803.
† Mochudi Farmers' Brigade, 'Nethouse horticulture in Botswana', *Appropriate Technology,* 1973, **3,** 15.

plant installation of adequate size (1 ha) to find out how to bring the capital cost down to a reasonable figure. About 5 mm of water can be evaporated in 24 hours, so that about half the desert area must be covered with stills in order to grow crops (with solar shields) on the other half.

6.8 FRESHWATER AND SEA FISH

Technology has provided sea fishermen and whalers with equipment so deadly that most of our favourite sea foods are becoming scarce. Trawlers are having to go farther and farther, and even with factory ships the energy ratio of energy in fish food to energy used to catch it is one-twentieth for UK fisheries and about one-hundredth for Adriatic fishing.* Almost all this energy is oil fuel for the ships, and it illustrates how dependent we are for our food on limited fossil fuel. British boats bring in less fish than they did in 1938†.

Marine fish are almost all carnivores, eating smaller fish, the base of the pyramid being microscopic plants, so it is unlikely that we shall ever find a way of increasing the fish in a given part of the sea or changing them to a more favoured species. All we can do is to regulate the catch, so that a species reaches equilibrium, and find tasty ways of using fish, and parts of fish, which are at present regarded as unpalatable. In the process of filleting fish the bones and intestines are of little value for human food, but at the same time about 30 per cent of first class meat of the fish, calculated on the weight of the fillet, is wasted. Good engineering could recover this‡. The use of parts of fish for animal feed is also a very inefficient process from the point of view of human food.

Pirie says that fish grow mainly on the continental shelf (sea less than 200 m deep) which has an area of 3×10^7 km^2—about that of Africa—and only about half of it is fully fished. Insufficient phosphorus is the limiting factor on fish growth, but all we can do is to make fuller use of all the areas where phosphorus-rich deep water comes up.

An important example of edible fish which is not caught, or, if caught, is used only for the inefficient process of animal feeding, arc the small shrimps (krill) that exist in huge amounts in the cold waters of the earth. As we have killed off most of the whales that used to feed on them, they are increasing rapidly. Now we shall have to develop processes ('the

* G. Leach, *Energy and Food Production*, 1975 (International Institute for Environment and Development, London).

† N. W. Pirie, *loc. cit.*, p. 159.

‡ O. Christensen, Food Technology Report, Fourth Scandinavian International Conference on Chemical Engineering, 1977, p. 12.

mechanical whale') which enable us to convert them into a tasty human food.

Fish farming in lakes and enclosed marine areas has often been considered, and the monks in the Middle Ages used to have their carp ponds which were fed by the privies*. Fish may be more efficient converters of food to flesh than land animals because they do not waste energy in body heat and do not have to resist gravity. The world production by aquaculture (5–6 M tons/year) is about 10 per cent of the total world fish catch, and is mainly carp in China. For the less developed countries fish could provide a supplementary animal protein, and the methods now being studied are the fertilization of warm water ponds to grown algae and vegetation on which a herbivorous fish or shell fish could feed. In India 1 ton of carp/ha–year can be produced without feed, while in Israel 2·5 tons/ha–year must be produced with supplementary feed to reach the economic break-even point. The main factors are the cost of land and labour.

The problem of using sewage sludge for growing fish in the developed countries arises from the presence of harmful pathogens—heavy metals, detergent residues, and persistent organic chemicals. The production of a biomass of worms and fly larvae at the sewage works, which can be fed to fish, is an interesting possibility: it has been claimed that a sewage works serving 10 000 people could produce 50 tons/year of biomass, and that this could produce 30 tons/year of live fish—an excellent example of recycling if it can be achieved†. Britain disposes of 1·4 M tons/year of dry sewage solids to the land or sea (1 M tons/year primary sludge, 25 per cent fat, 16 per cent average protein; 0·4 M tons/year biologically treated secondary sludge, with 40 per cent average protein).

It has been suggested‡ that Britain could find 1000 ha of offshore fish farming areas in which marine flatfish could be grown. Juveniles would be raised in a hatchery and the fish transferred to pens, kept above 15 °C by waste heat from power stations, where they would be fattened on fodder. This process could be economic at present because of the high value of such fish (£1000/ton), but is clearly not viable in the twenty-first century because better uses will be available for waste heat and for the fodder.

* This subject has been reviewed by C. J. Shepherd, *Jl R. Soc. Arts*, 1975, **123**, 807.

† 'Let them eat sewage down at the fish farm', *New Scientist*, 12 January 1978, p. 70.

‡ N. M. Kerr and K. T. Howard, 'The potential for marine cultivation in the United Kingdom', *Jl R. Soc. Arts*, November 1975, 811.

CHAPTER 7

The engineer's role in providing a worthwhile job for everyone

7.1 WHY PEOPLE NEED A WORTHWHILE JOB

The whole objective of engineers in the last 200 years since the Industrial Revolution has been to increase production per man hour and at the same time to develop new products which the industrial worker could purchase so that his rising standard of living enabled industry to continue to employ him. This has been very good up to a point, as has been discussed in the first three chapters, but it has had two disastrous consequences from the point of view of the human being employed in production. The first of these is *job dissatisfaction* or work alienation. The skill of the craftsman in the workshop and the peasant in the fields has been destroyed by jobs which are boring and repetitive. These modern jobs are broken down into small component parts so much that the human being has no pride in the quality of the finished product nor responsibility in the success in producing it. The second is the more subtle one of *unemployment* resulting from the fact that the exponential growth in production is eventually brought to a halt by the limitations in the supply of raw materials (especially fuel), space and consumer resistance.

Figure 7.1 is taken from a paper 'Beyond the Protestant Ethic' given by Ieuan Maddock*. It shows the change in the percentage of people employed in what are defined as primary, secondary, and tertiary activities. The proportion on the primary activities has fallen to a very small percentage. In Britain this is mainly farming, but it also includes mining, fishing and forestry. The secondary activities are what are normally called production processes in factories, which are mainly processing primary products into saleable goods, while the tertiary activities are the service activities concerned with transport, travel, distribution and shopkeeping, repairs, teaching, medical services, and the Civil and Armed Services; it also includes white collar workers in industry. The percentage engaged in secondary activities has remained fairly constant in the last

* British Association Symposium, 'Technology, choice and the future of work', see *New Scientist*, 23 November 1978, p. 592.

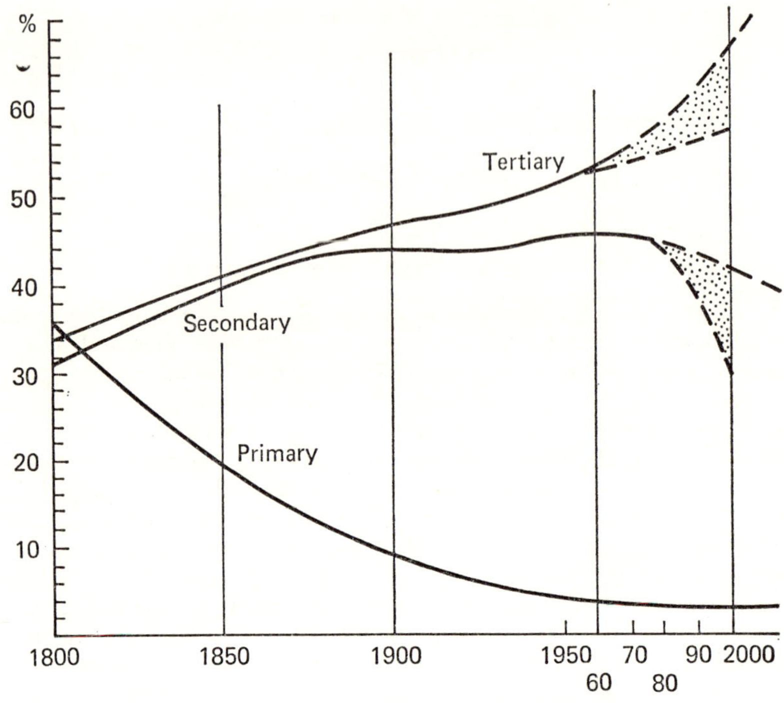

Fig. 7.1 *Changing pattern of employment*

100 years, but is predicted to fall severely in the future with the introduction of automation, the microprocessor, and robots.

In the long term there will inevitably be three trends which will still further reduce the number of people employed in the secondary activities.

(1) To save raw materials and make it possible for all humanity to have consumer goods, such as cars and refrigerators, these will have to be made to have a very long lifetime (fifty years or more). Built-in obsolescence and throw-away goods must come to be considered as disastrously short sighted.

(2) Limitations on energy, and particularly on natural hydrocarbons, must necessarily curb our tremendous expenditure on advertising, fashion, and unnecessary change. From this point of view the economy of the next century will be much more like a wartime economy.

(3) The colossal expenditure of human resources and raw materials on weapons production must come to a total halt if we are to reach a stable civilization.

Thus, the arguments of this book lead to the conclusion that, from the long term point of view, the number of man hours to be spent, in the future, on production will go down drastically. Many writers starting off from a more conventional forecast have reached a similar conclusion. John Fyfe* points out that unemployment levels have been rising in a continuous fashion throughout industrialized countries in recent years, and gives his opinion that the economic upturns of current and future cycles will no longer take up much slack in the labour market.

Ieuan Maddock (*loc. cit.*) has pointed out how the unemployment in the developed countries has shown a steady rise, with cyclic fluctuations imposed on it, ever since the Second World War, and now is at a figure of 5–8 per cent in all developed countries.

Philip Sadler† has said how the long term trends of the decline in employment in agriculture and the growth of white collar employment have continued, but that since the second half of the 1960s the decline in the proportion of employment accounted for by secondary industries has shown up. The main falls in employment in the UK have been in coal mining and in the older established production industries, such as construction, cutlery, cotton and wollen goods, motor cycles, and electrical machinery. There has also been a decline in two major service industries: transport and distribution. In his opinion it is necessary to go steadily ahead with automation, robots, and microprocessors, since, although these cause the loss of more jobs, they are necessary to retain international competitiveness, both with the more advanced developed countries and with the under-developed countries where labour is much cheaper.

Tom Stonier‡ in a forecast to the Government 'Think Tank' concludes that no more than 10 per cent of the labour force will be required to supply a technologically advanced society with its material needs.

In Chapter 3 I have taken it as axiomatic that *a stable civilization is impossible unless ordinary people have self-fulfilling lives.* If one pays somebody a dole for doing nothing all their life, however adequate this dole is in terms of purchasing power for a high standard of living, it is quite certain that this person will not have the feeling at the end of his life that he

* John Fyfe, 'Economics and employment', *New Scientist*, 23 November 1978, p. 597.

† Philip Sadler, 'Automation and Europe', *New Scientist*, 23 November 1978, p. 597.

‡ 'Professor shows how technology will change the society', *The Times Business News*, 13 November 1978.

has had self fulfilment. Schumacher* says: 'The Buddhist point of view takes the function of work to be at least three-fold: to give a man a chance to utilize and develop his faculties; to enable him to overcome his egocentredness by joining with other people in a common task; and to bring forth the goods and services needed for a becoming existence'. Lynda King-Taylor† has the following to say on the needs of people: 'People have very definite psychological needs, and demands for their satisfaction are being manifested in every industrialized nation. People need:

security and social order;
affection and love;
status and success;
dignity and self respect;
opportunities to learn-achieve-and-advance;
consideration as responsible and democratic people;
challenge and self esteem.'

A doctor at the British Association Conference on 'Technology, Choice and the Future of Work' (22 November 1978) said that unemployment actually kills people, and it is a well-known fact that some men die quite soon after retirement, and that there is a great reluctance by people to retire early to make room for others.

I should like to express the reasons why I feel that lack of a worthwhile job increases enormously the difficulties of a human being having a self-fulfilling life in terms of three factors.

(1) *It is a fundamental need of normal healthy human beings to feel that society needs their work and to take professional pride in doing a job well.* This is a very important activity of the human conscience. It is often denigrated these days with the label 'The Puritan Ethic'. However, it is certainly a necessary part of a human's self respect to feel that they are repaying society for all that they receive from it. The idea of bringing up a generation of young people who feel that society owes them a good living, regardless of anything they contribute, is as fundamentally evil as the idea that children can be educated without any effort on their part. The term 'workaholic' is used to assist in the destruction of the traditional values, but anyone, such as a mother who has brought up a family,

* E. F. Schumacher, *Small is Beautiful*, 1973 (Abacus, Tunbridge Wells, Kent) p. 45.
† L. King-Taylor, 'Creating an interest in work through participation', *Jl R. Soc. Arts*, June 1976, 380.

knows that at the end of the day there is far more satisfaction in a worthwhile job well done than in 'killing time'.

Herzberg says 'The hobby cannot give the complete sense of growth, the sense of striving towards a meaningful goal that can be found in one's life work . . . we cannot help but feel that the greatest fulfilment of man is to be found in activities that are meaningfully related to his own needs as well as to those of society'*.

(2) *Most human beings need an organized framework for their lives.* There are very few people who have so much self discipline that they can make themselves work productively, on any regular daily basis, for many years. Successful authors do this, but it takes considerable development of habit and will power to achieve it. In the rest of us the imposed discipline of set hours of work is desirable although once again, if one cites the housewife, 'a woman's work is never done'. The ideal would be to have a daily job, within walking distance of one's home, but this is rarely possible in an organized society. This does not necessarily mean that everybody needs a forty-hour week, fifty weeks in the year for forty-five years of their life, but most probably do require a framework of at least half this in a lifetime.

Meakin (*loc. cit.*, p. 184) quotes William Morris' horror at the idea of the leisure state. In the *Art of the People* Morris says: 'all we can do with it is to shorten that toil to the utmost, when the hours of leisure may be long beyond what man hopes for and what shall we do with the leisure if we say all toil is irksone. Shall we sleep it away and never wake up again in that case.' Meakin concludes the only hope for a future society is in daily labour as a moral value.

(3) *The satisfaction of human creativeness by using creative talents stretched to the full limits of their possibilities on an activity which is appreciated and of value to other people, is the highest form of self fulfilment available to the individual.* Creative hobbies can provide this to some extent and are, in any case, essential to make up for those aspects of creativeness which cannot be fulfilled by the job. Nevertheless, it is vitally important to find some way of replacing the job satisfaction destroyed by the engineer's development of sub-division of tasks, mass production, and the assembly line. Most unskilled jobs in modern factories are far less humanly satisfying than the jobs of: (i) a craftsman who is responsible for a whole individual object, (ii) a person who is concerned directly with producing food from the ground, and, perhaps above all, (iii) a person whose job is to help other

* David Meakin, *Man and Work*, 1976 (Methuen, London) p. 182.

people or deal with animals. It is another of our disastrous ideas that human service activities must be done by a machine wherever possible (e.g., the replacement of bus conductors by ticket machines) and to categorize a lot of service activities as menial instead of regarding them as opportunities for cheerful human relationships. It is absolutely essential for a stable society that the *educational system should encourage the development in the young of their creative talents, whatever they may be, and their ability to regard human relations as a vitally important part of life.*

Ieuan Maddock (*loc. cit.*) suggests that if industry goes forward in the present direction there will be four increasingly difficult problems.

(1) It may be necessary for a production worker to change job skill every ten years and perhaps even to move countries.

(2) There will be an increase in polarization of the work force into a few elite machine designers and maintainers, separated from an increasingly large, unskilled group.

(3) Whole industries will vanish.

(4) Work will become increasingly dehumanized.

It is quite clear that we cannot simply allow the automatic economic forces and short term political decisions to continue to deteriorate job satisfaction and increase unemployment. One can state categorically the following principle. *If we are to have a stable world society in the next century we have to find some system whereby every normal person is able to earn their living by a job which requires trained skills and gives them a feeling of responsibility for a worthwhile end result.*

The first clue as to how this can be produced is put very nicely in a paper by W. A. Dyson entitled 'Towards a new work and income orientation'*. He points out that there are two sectors of work which predate the three sectors dealt with by the economists, and which are never considered by them because money does not change hands. He calls these:

(1) The quaternary marginal mixed (cash and non-cash) labour sector. This covers hunters and fishermen who work and live on a survival and minimum life maintenance basis, non-economic farmers whose cash income is below the minimum cash level (these two categories are probably commoner in Canada than in Britain), and people who do odd jobs and services for direct cash payment, such as the handyman, charwoman, and baby sitter.

* W. A. Dyson, 'Towards a new work and income orientation'. Initial working paper by the Vanier Institute of the Family, Ottawa, Canada, 1 October 1976.

(2) Quintenary non-cash, familial and home (domestic) sector. That is all those persons who work full-time at home—child rearing and home care tasks. Their work is perhaps the most essential of all.

He states:

> That as our economic lives (household affairs) become less and less self, family and group reliant, to that extent do we become increasingly economically dependent and vulnerable.

and that among the routes towards 'economic health' are:

(1) the re-personalization of work, involving an increase of interpersonal 'economic' relationships,

(2) the reinforcement of living patterns that encourage an increase of self and group reliance for the production of goods, and the performance of services by oneself, within the home and family, and within the immediate community,

(3) the return of a large number of tasks now removed from the home and the immediate community.

> Support and legitimation to be given to those movements which are moving, in one way or another, with a focus towards home and community (e.g., home gardening, intermediate technology, 'do-it-yourself', neighbourhood day care, self-directed play, study and work, resource conservation, intra- and inter-community communication facilities, and so on extensively.

7.2 HOW CAN WE AFFORD A JOB FOR EVERYONE?

There are at least three conventional economic arguments which lead to the conclusion that it is impossible to offer a choice of interesting, worthwhile and reasonably paid jobs for everyone. These can be listed as follows.

(1) *The foreign competition argument.* This argument is that, unless we invest a lot of capital to replace people in manufacturing industries by sophisticated machines, our competitors in the highly industrialized countries will put us out of business in the export market, both in quality and price, so that we shall not be able to sell manufactured goods abroad and pay for our imports.

(2) *You cannot afford to pay craftsmen a proper wage because the machines can do it so much cheaper.* This is stated, for example, in a paper by Ernest Braun

of the Technology Policy Unit, Aston University*. He says that we must use micro-electronics both because, unless we raise productivity to the level of our international competitors, we shall have to leave the ranks of the industrialized countries, and because, unless we accept micro-electronics, the penalty is giving up a considerable array of goods, because some could not be produced at all on a craft basis (i.e., electronic equipment), and others would be extremely expensive because: 'it is salutary to think that crafts' products are generally speaking accessible only to those whose incomes are considerably above the incomes of the craftsmen'. This argument totally fails to take account of the fact that, in some products, the hand made craft article lasts at least ten times as long as the shoddy, mass-produced version, as well as giving much greater satisfaction and requiring a correspondingly smaller consumption of raw materials, especially energy. Goods of this kind produced by hand would undoubtedly cost more than mass produced ones because there might well be ten times as much labour, but they would not cost anything like ten times as much, because the capital investment in machinery is very much less, and because the materials although of much higher quality, do not cost anything like ten times as much, especially as the craftsman can make use of odd bits and pieces, whereas the mass production method requires identical brand new materials from which there is much greater wastage.

(3) *'I'm all right Jack, pull up the ladder'*. This expression categorizes the third group of problems. All demarcation disputes, attempts to preserve jobs which are making things which cannot be sold, or going slow in order to have plenty of overtime, come into this category, where the attempt is to preserve an industrial job which is no longer viable or prevent other people taking it over.

Clearly these obstacles are insuperable unless two conditions are satisfied, which at first sight appear totally impossible. They are:

(1) Economic decisions must be taken (not only in this country but in our rival countries) *on a long-term overall basis*. For example, account is not taken at present, when a manufacturer introduces a new machine, which displaces a workman, of the unemployment costs of this man because these are paid by a different authority. The employer purely calculates the economic cost of buying the machine, plus redundancy pay for the man displaced, and sets it against his salary. The national cost, including unemployment pay

* Ernest Braun, *Micro-electronics and Employment*, 1978 (Centre for Alternative Industrial Technology Systems, N.E. London Polytechnic, London).

for the man, is far higher; thus there will certainly be cases where the decision on a national basis would be quite different from that on a conventional employer's basis. In the long-term, therefore, the decision should be taken on a national basis, in other words, on whether: (i) the equilibrium society of the twenty-first century needs the product, and (ii) when full account is taken of unemployment costs it is still better to make it by the new machine or by the extra workman, plus existing machines. We can call this 'The Robot Fallacy', namely, the idea that it is always the most economic thing to do to replace a man by a robot, provided it saves money on a conventional calculation. As far as the international competition is concerned, so long as a country is competing with others which have not yet reached such a humane conclusion, it will have to subsidize the selected areas of production by the unemployment pay which is saved.

(2) The other essential condition for solution of these problems is that already outlined in Chapter 3, namely, that so long as a society judges the life success of an individual by how far their money and status symbols have gone above the average, there will continue to be a selfish scramble for an ever decreasing number of jobs. It has been pointed out, in the Vanier Institute report cited above, that money is not in fact the criterion of success, or even the motivating factor, in the quaternary and quinternary activities; for example, the housewife certainly does not judge her success in terms of money and yet does one of the most important of all jobs from the point of view of human life satisfaction (her own as well as that of her family).

Many people have thought about the same problem and arrived at the same conclusions. Schumacher's Buddhist Economics* is one example.

The Centre for Alternative Industrial Technology Systems (CAITS) which has been set up by the Shop Stewards' Committee of an aerospace firm at the North-East London Polytechnic† has been investigating the possibility of existing industrial firms concerned with very high technology (as for example in aerospace) turning to tasks of equally high technology but appropriate for the future, as for example, renewable energy or energy saving which can employ far more people than nuclear energy with a much smaller capital expenditure per person and at the same time

* E. F. Schumacher, *Small is Beautiful*, 1973 (Abacus, Tunbridge Wells, Kent) Chapter 4.
† Mike Cooley, 'New technologies, whose right to choose?', British Association Conference, Technology Choice and the Future of Work, 22 November 1978.

save scarce resources. A number of groups of shop stewards in other firms of high technology are similarly developing diversification proposals for their industry.

David Foster published a letter in *The Times** dealing with the problem of the unrest caused by unemployment, particularly of the young, and pointed out how the North Sea oil gives us strength on the international balance of payments but increases our unemployment. He proposed a three year National Service for young people under 25, with a year in industry and business, a year in agriculture and forest products, and a third year in social services, such as local government and the police. He proposed that these National Cadets would be paid a dignified living wage by the government so that their salary costs would not come on the employers who would, however, have the responsibility of training them to make good use of their service. He also proposed Vocation Colleges to assist in ensuring that the young people were trained to do work which was satisfying to them. This proposal was particularly concerned with making sure that school leavers had a worthwhile and well paid job as soon as they left school or higher education, as well as reducing unemployment and doing many worthwhile jobs which are uneconomic at present, particularly what he called 'orderly functions', that is 'those activities of a routine nature which tend to be neglected in the small business or on the farm'.

It is a strange fact that people would accept most of the changes required to ensure a job for everybody quite cheerfully under war conditions, but will not do so in peacetime. National Service for military purposes, which is certainly stultifying and in some cases depraving, in terms of teaching violence to the young, is very rarely accepted except in a war situation. There are several countries in which National Service of a constructive kind, such as manual work on road construction or working on the land, is already part of the education of the young. It would also obviously be highly desirable, if this period of National Service could be introduced, for it to include a period of training in the type of manual skills involved in 'do-it-yourself': making furniture, carpentry, and gardening, which we do not seem to be able to introduce sufficiently into the conventional educational system.

In my view, to solve the problem of providing an interesting and worthwhile job for everyone it is essential for the government to subsidize four areas of work which are needed in a stable society in the twenty-first century; these areas are as follows.

* David Foster, 'Unrest and affluence', letter to *The Times*, 30 August 1977.

(1) *Genuine human service.* At the top of the list comes teaching and health services. It is clearly nonsense to have unemployed teachers, and to close down teacher training colleges, because the number of youngsters is going down, when it costs nearly as much to pay these people unemployment pay as to pay them in their jobs. Obviously the right thing to do is to halve the size of the classes and have twice as many teachers, and this could certainly be done if the employers of the teachers were subsidized to the extent of the unemployment pay. The same thing applies to all the health services, especially nursing and medical ancillary services, and to the genuine social services of looking after helpless and elderly people. There are many other services which are genuinely valuable to human life but which are being cut down under the present short-term subdivided economic system, such as bus conductors, milkmen, policemen, home-helps, and even the helpful person behind the counter in a shop.

(2) *Agriculture.* At present the farmer is forced by the economic situation to go in for monocropping, cutting down hedges to make 100 acre fields, the extensive use of chemical fertilizer sprays, herbicides, and insecticides, and treating the soil as though it were a mine rather than a permanent possession. This has to be done in order to save the costs of labour under the existing economic system. If he were subsidized by the unemployment benefit he could afford to employ far more people and revert to the old established methods of crop rotations, mixed farming, and dealing with the weeds in the most effective way, that is by hand hoeing. Even if one only converted some fields on a farm back from monocropping cereals to kitchen gardens, which require much more extensive work by hand, the benefit would be very considerable from the point of view of food to the nation. There is little doubt that most human beings would prefer to do work connected with producing food from the land to being totally unemployed, especially if they did not have to work the very long hours of the traditional peasant. It would also obviously be desirable to have extra people working on the land at certain times of the year (e.g., sowing and harvesting) and far fewer in the winter.

(3) *Conservation engineering.* There are a great many industrial manufacturing tasks which are uneconomic under the present system, but which are necessary to the stable society of the twenty-first

century, and could be made economic by the unemployment subsidy. These will be discussed in full in section 7.3 below.

(4) *The return of the craftsman*. There are certain products which people need which are far more satisfactory and long lasting if made by hand by craftsmen than if made by machine. While it is possible that we would have to continue to have a certain proportion made by machine and selling at a price perhaps 15 or 20 per cent lower, it is clearly desirable to subsidize unemployed people to work in local workshops to make them by hand, using available raw materials. Varying the ratio of hand-made and machine-made could provide flexibility to make sure jobs are available for everyone. These areas include furniture of all kinds, clothes, and children's toys, and also the skilled maintenance and repair of old but well made existing objects. We could replace the throw-away society by the conservationist, craft-based society.

7.3 CONSERVATION ENGINEERING

In this section I shall try to give a comprehensive list of those activities related to engineering which satisfy the following criteria to make them suitable for an equilibrium society in the twenty-first century.

(1) They minimize the use of irreplaceable raw materials and damage to the environment.

(2) They provide relatively interesting and worthwhile jobs for as many people as possible.

(3) The capital investment per job is minimized so that they could be applied all over the world for all people.

7.3.1 *The elimination of pollution*

The causes of pollution from an industrialized society have already been discussed in previous chapters. Wherever possible pollution should be eliminated by improving or altering the actual process itself, for example, by perfecting combustion or going from the spark ignition engine to the diesel engine. Where this is not possible the cleaning of the effluent gases or water would be done in such a way that the pollutant would be recovered and available for recycling, and the same would be applied to all solid waste.

7.3.2 *Recycling*

It has long been customary to recycle iron, copper, and aluminium, and, indeed, in an equilibrium economy the amount of metal ore to be mined could be reduced to a very small proportion, almost all being recycled. However, there will be a considerable use of further raw materials during the period that the population of the underdeveloped countries achieves a fully adequate standard of living.

Plans have already been put into effect in some cases*. For this purpose it is necessary to separate domestic refuse into the following categories:

(1) paper,
(2) glass,
(3) tins,
(4) polythene film,
(5) plastic containers,
(6) textiles,
(7) aluminium.

Ultimately, no doubt it will be necessary for every household to have at least four separate bins, one for paper, one for textiles and plastics, one for metals and glass, and one for combustible or organic matter. This would reduce the separation at the centre enormously.

Organic compostible materials will be composted and returned to the land or else used in a first stage for methane generation, or the cellulose material will be converted, by hydrolysis, into ethanol†.

The dustman will be regarded, in such a society, as a highly skilled and respected craftsman, since all refuse will be regarded as valuable raw material, and its handling will have to be carried out with skill and reliability.

City night soil will always be used as a source of methane and also of phosphorus and potassium for return to the land. Methods will have to be developed for extracting heavy metals, particularly lead and zinc, from the sludge, and the soluble nitrogen from the liquid, so that these can be reused. The purified water can also be reused. The solid matter will have to be mechanically de-watered, as far as possible, on continuous filter presses, before return to the land. The sewage works will become a chemical plant operated under very precise control as a major source of recycled materials.

7.3.3 *Repair and modernization of houses and machines*

In an article by John Davis‡ a pilot study, by the Battelle Research Centre

* Jon Volger, 'Oxfam Waste Saver Project', *Proceedings of the Newcastle Conference on Alternative Technology*, 1976.

† A. Porteous, 'Hydrolysis of domestic refuse to fermentable sugars', *Consumer waste recovery and recycling*, April 1974 (Institution of Chemical Engineers, London), Paper A25.

‡ John Davis, 'Alternatives', *Guardian*, 2 November 1977.

in Geneva for the EEC, on the the potential for substituting manpower for energy was described. In this it is pointed out that if the useful life of a car is doubled from ten years to twenty, the energy and material consumption, and the labour in the manufacture, would all be nearly halved, but the extra labour required in service stations, garages and reconditioning workshops would substantially exceed the number of jobs lost in the manufacturing industry. Every mechanic knows that repair and maintenance is much more interesting work than mass production, because every vehicle requires individual diagnosis and treatment. The same applies to all other consumer machines, such as clothes washers, dish washers, vacuum cleaners, refrigerators, and deep freezers, and it is quite certain that none of these could be made available all over the world unless they are made to last a very long time.

This means that the elimination of built-in obsolescence not only saves raw materials and energy, but also increases the number of jobs and their human interest. Spare parts and repair manuals should be readily available. Exactly similar considerations apply to housing. The consumption of raw materials—steel, cement bricks, glass—and energy for new houses is enormously greater than the corresponding consumption for modernizing and improving existing houses, whereas the number of jobs of an interesting character is much greater in the modernizing process*. The ventilation and draught proofing of existing buildings to save a big fraction of the heating energy is a very important aspect.

7.3.4 *Fuel economy*

This has already been discussed in detail in Chapter 4, and all aspects of fuel economy provide plenty of interesting and worthwhile jobs, and are extremely important from the point of view of reducing the consumption of the one *non-recyclable mineral*: fossil fuel. A number of examples of fuel economy are good examples of high technology, combined with work of long-term importance. In this area we have the development of telechiric coal mining and of solar and wind power. The House of Commons Select Committee on Science and Technology† has concluded that in Britain we could obtain 15 MTCE/year from solar power by the year 2000, which is more than we get currently from nuclear power. Photovoltaic technology is very sophisticated and the proportion of jobs, per

* D. Elliot ('Can alternative technology create jobs?', CAITS paper) points out that in Britain at the moment there are a quarter of a million unemployed construction workers and a massive national shortage of decent housing.

† Third Report of the House of Commons Select Committee on Science and Technology, Session 1976/77, Vol. 1, p. Li.

unit of energy, in this field has been calculated to be about 6·6 times higher than that in nuclear energy, and the capital cost would be less than half as much*.

The chemical engineering processes involved in the production of ethanol from agricultural cellulose (e.g., straw), and from the cellulose of domestic refuse, is another area where advanced technology research is of great potential benefit to humanity. In the case of agricultural refuse, it would be particularly helpful if processes could be developed whereby the transport of the straw is reduced to a minimum, for example, some relatively low capital cost plant for use on each farm.

The generation and storage of power from wind is also an area in which a great deal of high technology research is already being done and it is very suitable work for aerospace firms. It has been estimated to have a $2\frac{1}{2}$ times better labour/capital ratio than nuclear power. This will be especially valuable if relatively small units can be developed, of the order of 10–100 kW, for village power, since these could readily be installed all over the world wherever there is a reasonably high average wind flow.

The conversion of existing fossil fuel fired powered stations to the passout steam cycle, and the use of coal, with fluidized bed and very good gas cleaning processes, in these stations is also an area where a comparatively high labour to capital ratio, and very interesting work, can be obtained for the developed countries.

The last example where fuel economy can be combined with more jobs and more interesting work is that of transport. Various ideas have been discussed in Chapter 5, such as the development of very reliable and convenient public transport, the hybrid or fuel cell car, and the high efficiency coal-fired steam locomotive. Post buses for village transport is a worthwhile example.

7.3.5 *New materials*

A great deal of work on the development of new materials has been done in connection with military and space purposes, and has shown clearly how, if we deliberately set out to develop improved materials of the types required by the equilibrium society of the twenty-first century, we could make tremendous improvements. Particularly valuable would be improvements in regard to corrosion resistance, fatigue resistance, and other properties required to give the possibility of making consumer durables really durable, that is, to last some fifty years with relatively few replace-

* *Jobs from the Sun*, 1978 (California Public Policy Center, USA).

ments and not too much maintenance. In the last century kitchen utensils lasted a lifetime.

7.3.6 *Safety engineering*

This is an area of vital importance where the subsidy derived from not paying redundancy pay and from the higher prices that become worth paying for objects that last many times as long, will be employed to give a much higher level of safety than we have had in the age of planned obsolescence. This applies to all kinds of accidents which could be averted by more careful engineering design in the home, in transport, in the factory, in the mine, and on the farm.

An example quoted in the *New Scientist** is the electromagnetic retarder braking system for road transport vehicles. This is fitted to the propellor shaft and when battery current is fed through the stator it reinforces the brakes and reduces overheating. Such a system has been compulsory for coaches in the mountainous terrain in France since the mid-1950s.

7.3.7 *Medical engineering*

This subject will be discussed in the next chapter and it undoubtedly is an area where our present topsy-turvy economic priorities would be totally changed in an equilibrium society where a choice of jobs is available for everyone. One example is the kidney machine; Mike Cooley† describes how 3000 people die each year in Britain because they cannot have access to a kidney machine, and considers that this should be one area that should be immediately subsidized by the dole money otherwise needed by the extra people employed. Our work on carriages for handicapped children at Queen Mary College Mechanical Engineering Department has shown a tremendous demand which can only be satisfied at present when a charity comes forward with the necessary money, or by free labour.

* Lucas Aerospace Combined Shop Stewards' Committee, 'Dole queues or useful projects', *New Scientist*, 3 July 1975, p. 10.

† Mike Cooley, 'New technologies: whose right to choose?', British Association Symposium, Technology, Choice, and the Future of Work, 22 November 1978.

CHAPTER 8

Medical engineering

8.1 DOCTORS AND ENGINEERS

Not unnaturally many people look up to the doctor as someone dedicated to the improvement of human life and, correspondingly, despise the engineer as someone whose work frequently causes a deterioration in human life, by side effects causing inconvenience or discomforts, or even by weaponry. Many people who have studied the matter in depth, such as Ivan Illich*, point out that the medical profession is in fact just as responsible as the engineering profession for the many illnesses and mental depressions to which those living in modern cities are subjected. This is because the doctors have concentrated on curative medicine, especially the use of a whole new range of drugs and antibiotics, instead of applying the greater part of their skill to preventive medicine. The engineer must accept responsibility for many of the insidious poisons, such as lead, carbon monoxide, sulphuric acid, and radioactivity in the air, and DDT in our food, but it is surely the job of the doctor to sound the warning against these poisons, and against messing about with natural food, taking out good constituents, which are then sold separately as medicines, and adding all kinds of preservatives. The engineer provides the motor car which robs us of walking exercise, and the doctor merely treats the arthritis which results at the end of our lives, or gives us pills to enable us to sleep when we have not had enough exercise. It is clear that *we can only create a situation in which people's minds and bodies have a chance to develop to their full potential by a joint responsibility of the doctors and the engineers.* It is, unfortunately, true that, until very recently, the relationship in the hospital between the doctor and the engineer, or the surgeon and the engineer, was essentially that of master and servant, that is, the doctor or surgeon was head of a team, and the engineer was only called upon to tidy up and construct the ideas designed by the medical man. There is little doubt as to what is the only way to real success, both in regard to preventive medicine and healthy life, and in engineering contributions to medical and surgical repair of damaged bodies. *The engineer can only*

* Ivan Illich, *Medical Nemesis*, 1975 (Calder and Boyars, London). He says (p. 11.): 'The medical establishment has become a major threat to health'.

make a really effective contribution if he is in an equal partnership with the medical people. There is a move in this direction. For example, in 1976, at the Biomedical Research and Development Unit at Roehampton in London, an engineer replaced a doctor as director. This institute works primarily on prosthetics and orthotics for amputated or damaged legs.

Many engineers have felt strongly impelled by their conscience to work in this field and a great deal of good work has been done as a result. However, in general, very much more support is available for conventional, purely medical projects, especially the development of new drugs, and as far as engineering is concerned vastly more money is available for high technology and military work. In a document entitled 'To choose a future' prepared by the Swedish Royal Ministry for Foreign Affairs, it is pointed out that there is a pronounced 'skew' in welfare planning in the sense that on-going research and development caters very well to certain problems, whereas vast needs for research efforts have remained completely ignored in other cases.

The engineer who works in this field would be well advised to read Illich's *Medical Nemesis* to get his objectives right. He writes (p. 169):

> Healthy people are those who live in healthy homes on a healthy diet; in an environment equally fit for birth, growth, work, healing and dying; sustained by a culture which enhances the conscious acceptance of limits to population, of ageing, of incomplete recovery and ever imminent death. Healthy people need no bureaucratic interference to mate, give birth, share the human conditions and die.

In this chapter I shall first of all consider the engineer's contribution to a more healthy life style and to preventive medicine, and then deal with a number of examples of medical engineering in which the engineer is already contributing to the achievement of a more natural life style for people whose bodies are damaged by illness or heredity.

8.2 THE ENGINEER'S CONTRIBUTION TO A HEALTHY LIFE STYLE

In the developed countries the engineer has already made the major contribution in the achievement of sanitation and hygiene in all aspects of human life. In some cases this has been done at such a cost in raw materials, fresh water, or energy, that it will not be possible to provide similar standards by the same methods for 8000 M people in the next century. Moreover, it has not been done with a view to recycling materials for plant fertilizers, as discussed in Chapter 7, so that there is much work for the engineer to do in terms of maintaining the present high levels of biological control with a much less spendthrift consumption. In addition, starting from the other end of the spectrum, it is urgently necessary to provide those people living in highly unhygienic and insanitary conditions, in many parts of the world, with a certain modicum of disease control, using as simple and as easily obtainable raw materials as possible, and without harmful side effects.

In regard to the chemical pollution of air, land, and water, and the effects on human beings of noise and accidents due to cheap engineering, the engineer must accept an absolute responsibility for bringing down all these harmful influences far below the level at which they can affect even the most sensitive person when subjected to them for a whole lifetime, or have effects on new-born babies at the lowest detectable statistical level.

The lack of exercise, from which most of our population suffers severely, while it is a consequence of the engineer's development of the motor car, can only be put right by a strong public realization of the ill effects on the human body from *failing to use it in the way for which it is designed.* Nevertheless, the engineer's work to improve bicycles and to develop a lightweight, two-seater tricycle with weather protection, might be one way of restoring human exercise as part of the normal daily journey.

Moreover, the main contribution the engineer can make to preventive medicine is to give a great deal more thought to safety devices to prevent accidents. There has been a school of thought in the USA which has regarded the body of the motor car as an extended suit of armour round the passenger so that he would not be damaged during the inevitable crash. This view has been strongly denounced by Ralph Nader who has stressed that *the car should be designed to make it very much less likely to have accidents*, and this is surely the engineer's first duty. Only a fraction of accidents can, however, be blamed purely on vehicle design; many more can be blamed on the fact that cars are designed, and sold, on the basis of tremendous acceleration for overtaking and very high top speeds. There will, however, always be accidents as long as there are a large number of individually driven vehicles competing for limited road space and subjected to human error and misjudgements, variable road surface conditions, and weather conditions such as fog. The engineer has, therefore, already developed safety belts and head rests that prevent 'head flip' when the vehicle is struck from behind (there are a number of people in Australia who are totally paralysed as a result of the damage to the spine and neck from such accidents). More advanced engineering designs, with foam rubber devices on the dashboard and steering wheel, and with pneumatic cushions that are inflated in front of the driver and passengers so rapidly in a crash that they prevent them hitting the rigid parts of the car, have been developed, and cars are now being fitted with devices so that they cannot be driven unless all seat belts are fastened, but this appears to me to be going too far in the direction of the armoured knight.

The motor bicyclist runs an exceptionally high risk of accidental damage, partly due to the vulnerability of the rider of a high speed two-wheeled vehicle, and partly due to the attitude of mind that usually goes with such a vehicle. It is doubtful whether high speed motor bicycles serve any necessary role, but the low speed motor cycle or moped does provide a very economical means of travel at a speed quite fast enough for longer commuting journeys (e.g., 25 miles/hr) which may be too far for the bicycle.

In the factory the engineer has already given consideration to safety devices of all kinds after accidents have occurred. *Once telechirics is fully developed it will no longer be necessary for any person to work in dangerous zones in the mine, quarry, undersea, or factory.*

In the home the engineer has probably not yet given enough thought to ensuring the safety of all the devices provided, especially those involving electricity, gas, or boiling fluids. This is largely the result of the highly competitive market in these fields, but it may also often occur that they are designed by an engineer without sufficient consideration of the needs of the housewife, and with the disastrous idea of designing for a short life to keep the market going.

8.3 ENGINEERING AND SURGERY

8.3.1 *Joint replacement and other implants*

A very considerable amount of excellent engineering research and development has been done to provide the surgeons with replacements for worn out and damaged parts of the body, particularly artificial hip joints and, more recently, knee joints, mainly for chronically disabled arthritic patients. S. A. V. Swanson and M. A. R. Freeman have recently published a book* which illustrates excellent cooperation between the orthopaedic surgeon and engineers. This is mainly concerned with replacement of the 'ball and socket' hip joint, by use of implants of cobalt chromium alloy and special plastics. There are many problems which have been worked on, with some success, in connection with the

* S. A. V. Swanson and M. A. R. Freeman, *The Scientific Basis of Joint Replacement*, 1978 (Pitman Medical Publishing, Tunbridge Wells, Kent).

fastening of the implants to the remaining bone. The knee joint is very much more complicated in its bone formation and movement geometry than the hip joint, but some successful replacements have been developed and are described by D. A. Sonstegard, Larry S. Matthews and Herbert Kaufer*. In normal activities, such as running, walking, kneeling, and climbing stairs, the load put on the human knee joint can exceed five times the weight of the body; games and sports can increase this considerably further. Eight thousand total hip replacements and 30 000 knee replacements were performed in the USA in 1966, and, if a more satisfactory knee replacement were available, surgeons would do the knee replacement very much more often. From the patient's point of view, knee or hip joint replacement, if successful, can give a great relief from pain, and increased mobility. While engineers can use their skill to solve these problems, as requested by the surgeons, this is surely one of the most clear cases where preventive medicine or even an extension of the non-medically recognized treatments, such as osteopathy, massage, and diet alteration at an earlier stage in life, might lead to a much more satisfactory life from the point of view of the patient.

After the surgeon has amputated a leg, at various points from the ankle to the hip, or an arm, he needs cooperation from the engineer to provide a prosthetic mechanical replacement. Here again the engineer has already made enormous improvements on the original wooden legs and hook forearms. One important problem is the sensitivity of the stump, in the case of a leg amputation, to the weight of the body carried through the artificial leg. The use of pneumatic supports for immediate post-amputation mobility, the use of a load sensing device on the sole connected to an audible signal to help the amputee to re-acquire walking skills, and the use of bacterially protected gaseous environment in a plastic bag sealed to the amputee's limb by a higher pressure plastic toroid, are all discussed in the 1977 report of the Biomedical Research and Development Unit at Roehampton. The engineer has made very substantial contributions to the design of unpowered artificial legs for above knee amputations, and very much work has been done on artificial hands controlled from other muscles and powered by battery or carbon dioxide.

Work is now being done on the replacement of human bones and teeth, by implants of biologically active materials†. Ceramic materials have been developed which can form an intimate bond with normal bone and may lead to the implanting of false teeth with artificial roots.

* D. A. Sonstegard, Larry S. Matthews and Herbert Kaufer, 'The surgical replacement of the human knee joint', *Scientific American*, January 1978, **238,** 44.

† C. R. Hassle, 'Bioactive teeth and bone implants', *Physics in Technology*, 1978, **9,** 272.

The engineer has cooperated with the surgeon in work on the human heart for many years. The simplest example was the replacement of a damaged tricuspid valve by a ball and socket valve, using a ball rather like a ping-pong ball as the valve. The cardiac pacemaker has been inserted in many patients to give electric stimulation to regulate the heart beats of a patient in whom they have become too erratic. The first ones had batteries and had to be replaced by surgery when the battery ran down. Another type has been developed which can be re-charged by electro-magnetic induction through the chest wall without surgery, and isotope powered cardiac pacemakers have been developed by the UKAEA* in which the heat generated by Pu^{238} (half life ten years) produces electricity in a thermo-pile, so that a life greater than ten years should be feasible. It is stated that with reliable canning, this β emitting isotope can be ultra safe, both to the patient and to the general public. The canning of the plutonium must remain intact under all conditions, including cremation.

Much work has been done, especially by chemical engineers, on the development of artificial kidneys. The human kidney is a remarkably well developed apparatus for removing all kinds of poisons from the blood stream by osmosis through an array of fine tubes with enormous surface area. At present the artificial kidneys occupy volumes some 100 times as much as that of the human kidney, and the patient has to have their blood circulated through the artificial kidney once a week by means of tubes pushed through the skin. Both in the case of the kidney and the heart it is possible that the engineer could ultimately develop an entirely synthetic substitute which could be inserted in the body. However, the problems of compatibility, which have already arisen in relation to heart transplants, make the ultimate human value of such a line of research very doubtful.

8.3.2 *Equipment for surgeons*

The engineer has much contribution to make to the improvement of surgical tools. Lasers are now being used in surgery†. Continuous argon lasers are being used in optical surgery for repair of retina detachment and for treating various diseases. Lasers have been used for treatment of skin lesions and to heal tropical ulcers, using low power density in the lasers, where the main result is the stimulation of healthy tissue growth. They have also been used to treat skin carcinoma, and at high power for

* UKAEA leaflet, London, 1969.

† D. W. Goodwin, *Physics in Technology*, 1978, **9**, 248.

incision and cauterization. There is no doubt that the laser is providing a valuable new tool for surgery.

We have worked with an Oxford surgeon in the development of a sewing machine device whereby pressing a trigger on a pistol-like piece of equipment pushes a needle through the tissue and transfers it to a second grip automatically, after which the pistol could be pulled to draw the thread through. This could be extended to develop hand sewing machines for surgeons which would put the stitch right through and complete a closed loop, for example by sealing the thread ends together thermally. We have also constructed, for another surgeon, a foot control device whereby he could operate very fine hand-held forceps and scissors without the movement of the hand which would otherwise be necessary to work them. The foot control had a virtual bearing passing through the ankle so that no movement of the leg was necessary when operating it. This control went to the implement via a long Bowden cable of the type used for camera shutters. A special table, with a half circular hole in one side was constructed so that the surgeon could sit down, and rest his forearms on this, manipulating the tools with his wrists and working through a low powered binocular microscope.

8.3.3 *Telechiric surgery*

There is no doubt that once the technology of telechirics has been fully developed it will enable surgery to be carried out with only the patient in the operating chamber and the operation performed by a series of pairs of telechiric stainless steel hands controlled by the surgeon from a panel outside the operating chamber, with full visual and tactile feedback (Fig. 8.1). There are many advantages of such a system once it has been developed.

(1) It is very much easier to ensure complete sterility of the operating chamber and instruments.

(2) One surgeon can control several successive pairs of hands, putting one pair into position, clamping whatever is necessary, and then moving on to another pair to handle the instruments. He can have his full manipulative skill applied in turn to each pair.

(3) The system can be applied to micro-surgery with magnification up to ten-fold, using miniature hands so that the surgeon feels and sees as though he were operating on a patient enlarged ten times. It is also possible to use fibre optic visualization to see round the back of what is being operated on. Any number of people can see

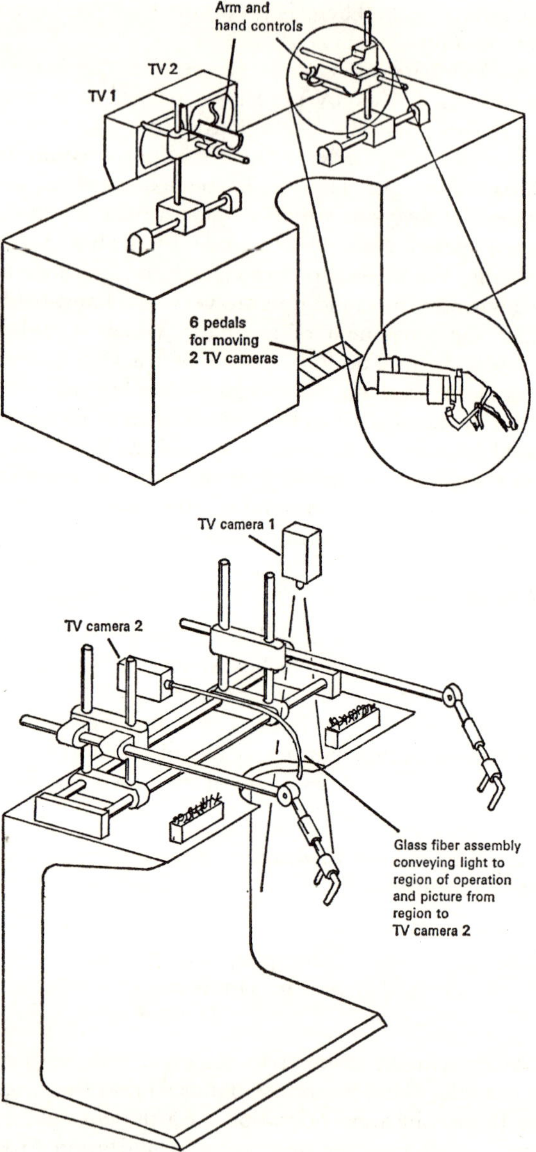

Fig. 8.1 *Diagram of telechiric surgery*

exactly what the surgeon is seeing on their own screens and assistants can operate other telechiric hands.

(4) The hands can be made of very slender stainless steel so that they require much less space than the surgeon's hands in spite of having greater strength.

It is also possible for a surgeon to carry out an operation in this way, without physically going to a remote site, provided the necessary skilled assistance is available locally. This could be very valuable in areas far away from a main hospital, e.g., lumber camps or oil rigs.

8.4 SCEPTROLOGY

The Greek word sceptre referred not only to the king's staff of power, but also to the crutch of the cripple. The word *sceptrology* has been coined, therefore, to describe *the science and technology of crutches*, using the word 'crutch' in its most general sense for any aid for people whose limbs are inadequate for their ordinary life tasks, which is external to the body. This section is not by any means comprehensive of all the work in sceptrology, but a number of devices in this field, particularly those which have been worked on in the author's Design and Inventions Laboratory, will be described.

8.4.1 *Battery powered stair climbing wheelchair*

I shall start with this problem as it is one on which I have been making prototypes for more than ten years. The staircase is a very ingenious way of making use of the human walking system to climb an inclined plane without sliding down it. By providing horizontal treads there is no need to use friction to overcome the pull of gravity and it only has to be used to provide the force for the forward motion as in walking on the level. Early work on systems with feet which were placed alternately on successive steps showed that this would require a very sophisticated control system since it had to adjust the rise and forward movement to the dimensions of the steps.

After trying a number of different wheeled systems, I reached the conclusion that a rimless wheel, as shown in Plate VI, could provide the ability, like that of a human foot, to put the lifting force entirely on to the horizontal surface. By having front and rear wheel drive the system could be made to climb staircases up to the maximum angle permitted in ordinary buildings, namely 37°. A rimless wheel gives a bouncing motion when running on level ground and it was found that by springing the spokes this could be greatly reduced, but not eliminated. The greater the number of spokes, the less the bouncing action, but on the other hand the greater the expense and weight. Thus the optimum number of spokes was a compromise, and the number twelve was chosen. It would be possible to have eight sprung spokes, each in the form of a trident, to give effectively twenty-four spokes. Another idea which was tried was to have soft tyres on the rear wheels and then have rigid spokes alongside the tyre which would come into contact with the step when the tyre was deformed by the edge of a step. Experiments showed that it was necessary to spring the spokes even in this case, but the springs would no longer have to be strong enough to carry the full weight, provided the sprung spoke was rigid against the large tangential force which occurs when it is lifted on to a step.

Plate VII shows a stair climbing wheelchair, on these principles, climbing stairs. This model, however, was incomplete in regard to the steering and the balancing of the passenger when the carriage tilts to go up or down stairs. If one wishes to steer up a staircase with a bend in it the spiral of the steps does not fit a four wheel carriage and it is necessary, either to spring the system very considerably to deform from a plane, or to have a three wheeled system.

Plate VIII shows the chassis of a system designed with three wheels. The front wheel would have to be driven at the arithmetic mean speed of the two rear wheels, and by means of the rollers on the spokes of the front wheel, the machine can be steered about the mid-point of the back axle by running one rear wheel forwards and one backwards. It is necessary to have a mechanical geared, two-speed drive to each wheel, because one requires very high torque on low rev/min for climbing stairs, and electric speed controllers do not give this ability.

The problem of levelling the person's seat requires that, as the carriage tilts to go either up stairs or down stairs, the seat moves upwards to put the centre of gravity of the person towards the upper wheel support to ensure stability against falling downstairs. This could be done, for example, by pivoting the seat about a point low down on the carriage, but since it involves lifting, it requires a powered movement.

It is hoped very much that we shall be able to complete this work, because there is undoubtedly a big demand for such a carriage in which people with crippled legs could go out to the shops and move about anywhere that normal people can walk. It might be necessary to have a system which could be extended in length for outdoor work, and for stair climbing, but shortened to take up minimum space when moving within a room. This is clearly a problem which could be solved completely, and

Plate I. *Three-dimensional diagram of the predicament of our civilization, as shown in Fig. 2.7. Either we climb laboriously out of the forest or we slip down to disaster.*

Plate II. An ancient symbol for conscience. *The Woman represents predominant Love, the Bull's trunk determination to carry out indefatigable labours, the Lion's legs courage and faith in one's 'might', the wings of the Eagle meditation 'on questions not related to the direct manifestations required for ordinary being-existence' All and Everything, p.* 310.

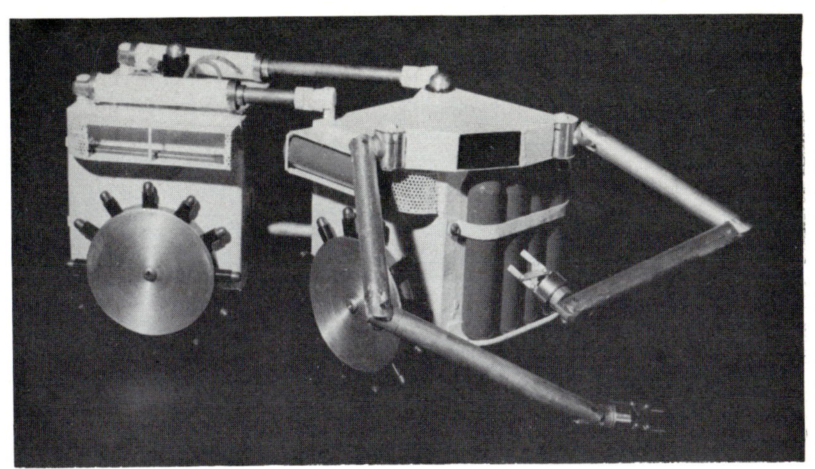

Plate III. *Model of telechiric mining device.*

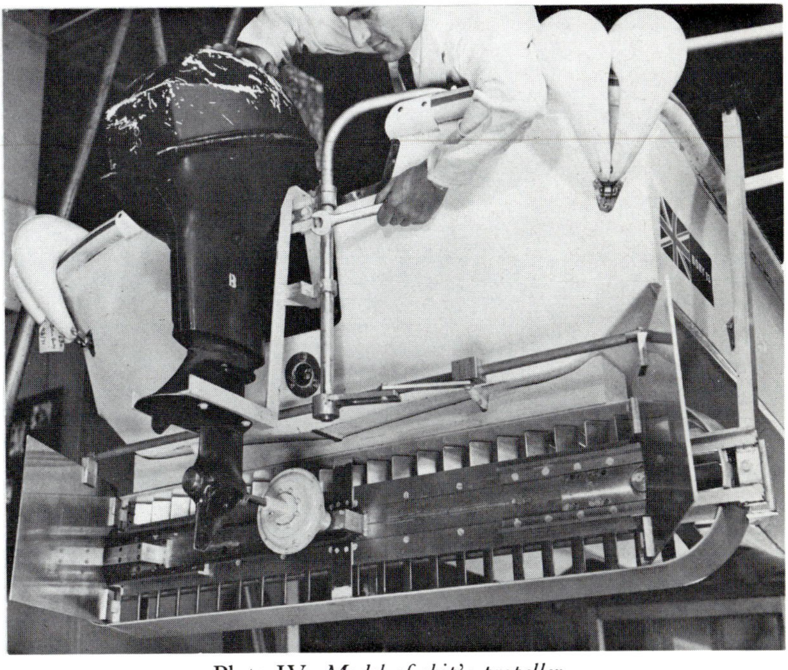

Plate IV. *Model of ship's propeller.*

Plate V. *Walking tractor.*

Plate VI. *Rimless wheel.*

Plate VII. *Stair-climbing wheelchair.*

Plate VIII. *Chassis of three-wheel stair-climbing device.*

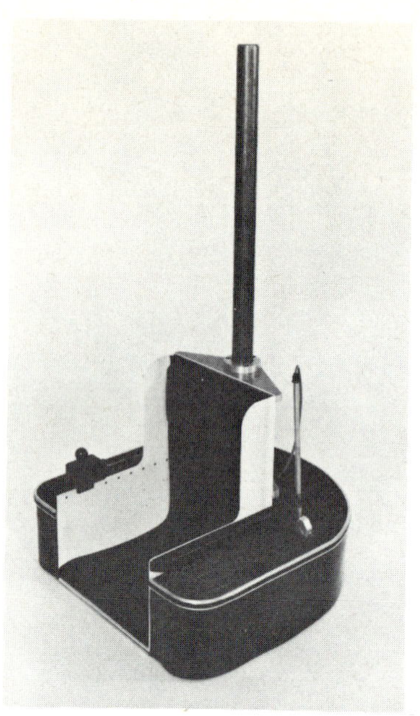

Plate IX. *Carriage for thalidomide children.*

Plate X. *Go-kart for spina bifida children.*

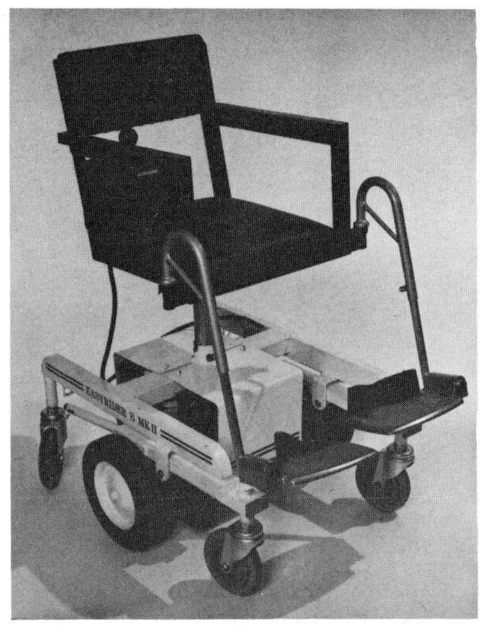

Plate XI. *Powered wheelchair for adults.*

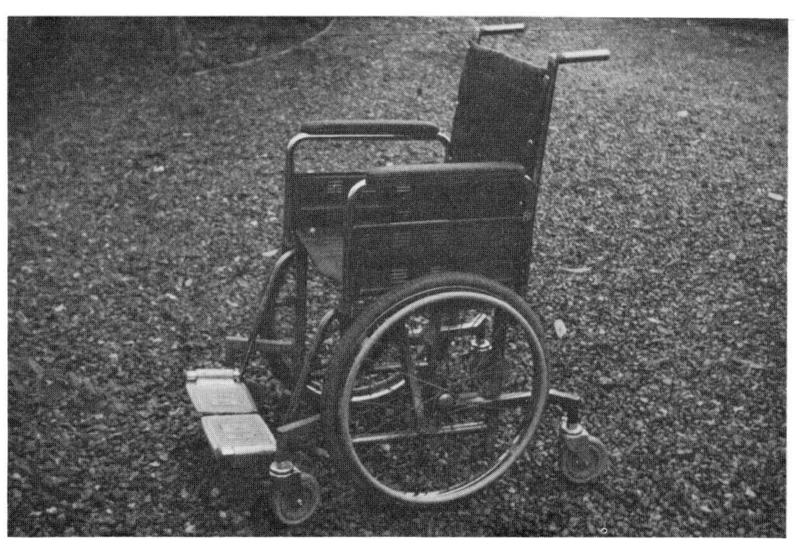

Plate XII. *Hand-operated wheelchair for adults.*

Plate XIII. *Stair aid.*

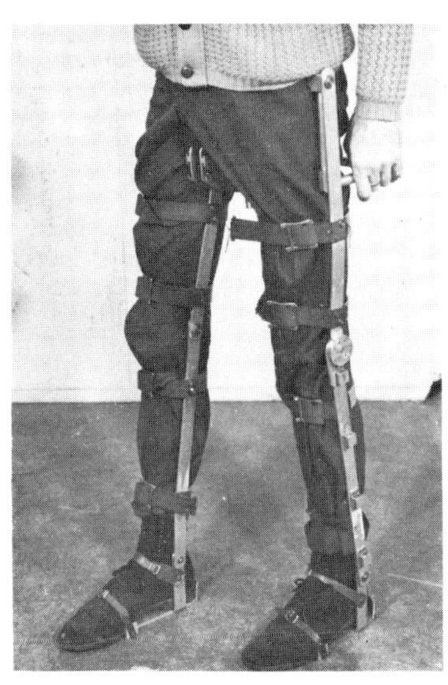

Plate XIV. *Exoskeleton.*

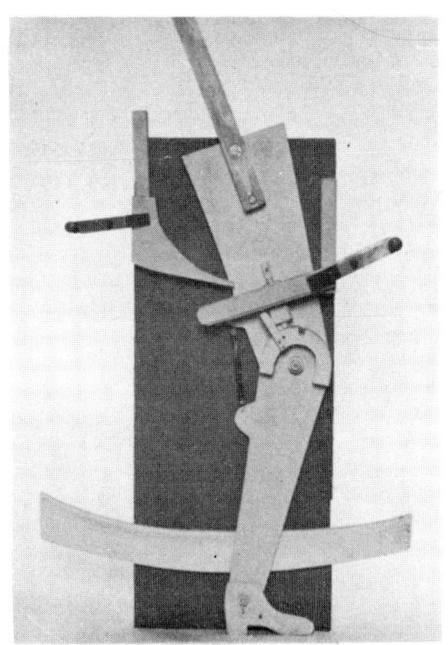

Plate XV. *Powered exoskeleton.*

Plate XVI. *Tea pourer for disabled.*

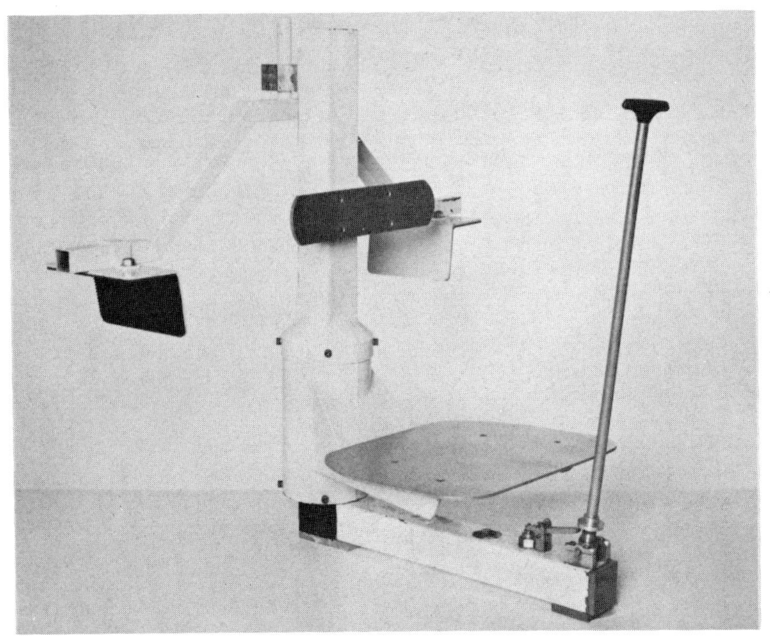

Plate XVII. *Aid for getting in and out of bath.*

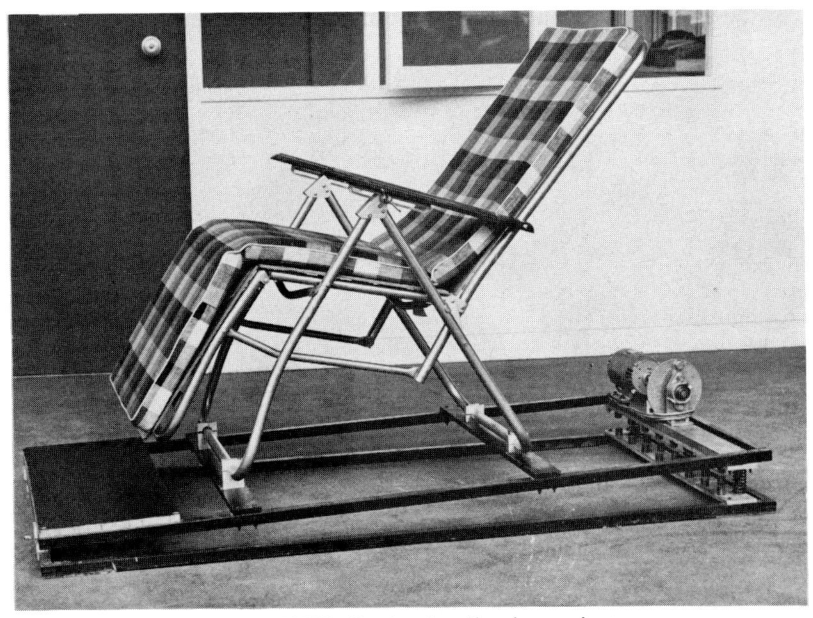

Plate XVIII. *Device for vibrating patient.*

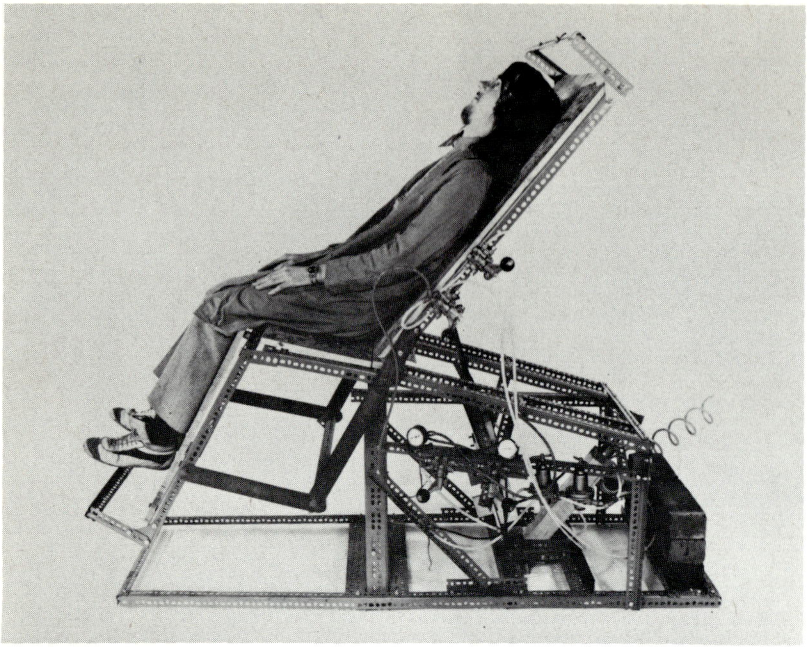

a

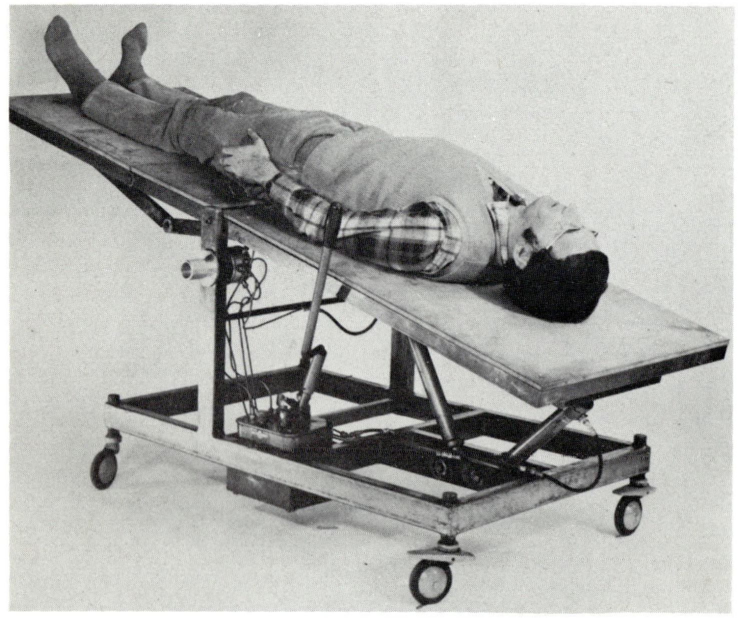

b

Plate XIX. *Fully variable hospital bed design.*

with a very high level of safety and reliability, in less than a year if it were given the kind of funding and resource priority given to the development of a new weapon of war or to a project connected with nuclear energy.

8.4.2 *Improved mobility for cripples on comparatively level ground*

The Medical Engineering Projects Group, led by Mr Eric Booth, in my Design and Inventions Laboratory, have developed, over the last ten years, a whole series of powered mobility aids for children and adults who cannot use their legs. Early on a system designed by Mr K. Shepherd was developed in which two separate battery-driven electric motors drove wheels, and there were castors of fairly large diameter, back and front, forming a diamond pattern. By reversing one motor and running the other forwards it could spin about the mid point, and it could go in any direction, forwards or backwards, by driving the two motors forwards or backwards at various speeds. This carriage was also developed with a lifting device to bring thalidomide children up to normal height (Plate IX).

Although this system was very compact and very mobile and stable, it suffered from the disadvantage that if the front castor rose on to a thick carpet, or there was a move from level ground to a gently sloping ramp, the two drive wheels were lifted off the ground because they were on a plane defined by the front and rear castor. Mr Booth invented a very satisfactory solution to overcome this problem by setting the front wheel and the two central drive wheels on a bogie which was attached to the main frame (which carried the rear wheels) by a horizontal bearing. By suitably positioning all the points of support and the bearings, it was possible to arrange that most of the weight was taken on the drive wheels in the centre, under all conditions, provided there was a stop to limit the ultimate movement of the bogie to prevent falling forward or backward. This principle has been applied also to hand operated wheelchairs to enable them to climb kerbs of height such that the driving wheels can lift on to them and to many other types of carriages since it combines the stability and mobility of a pair of reversible central driving wheels with the adaptability of a system with only three wheels.

Plate X shows an electrically powered go-kart for spina bifida children in their open air playground. Plate XI is a powered wheelchair for adults, and Plate XII is a hand operated wheelchair for adults, all based on this principle. In this connection studies have also been made on the ergonomics of wheelchairs. It has been found that the conventional collapsible seat made of a canvas sheet supported at the two sides is not a comfortable seat because it sags towards the middle. The ideal seat is a pair of round concave surfaces supporting the ischial tuberosities.

8.4.3 *The wheelchair user who drives his own car*

Figure 8.2 shows a design prepared for a device whereby the user of a collapsible wheelchair, who has full strength arms, can get into his own

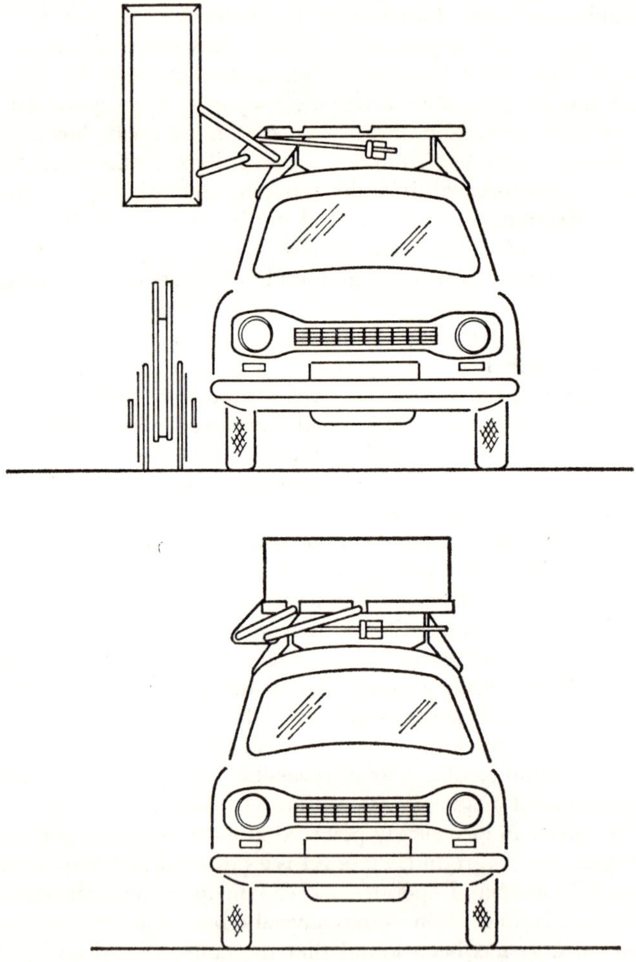

Fig. 8.2 *Device for stowing a wheelchair on the roof of a car*

car from his wheelchair, collapse the wheelchair, and stow it on the roof of the car so that he becomes completely mobile without any assistance.

8.4.4 *Mechanical aid for stair climbing*

Plate XIII shows a device which has been developed, at the suggestion of a chest surgeon, to assist people who cannot climb stairs owing to heart or lung problems. The device is very much cheaper and simpler to install than a lift and consists of a powered bar, with a microswitch on the bar, so that the weight of the person is taken upstairs, and they have to do no more work than if they were walking on level ground. The microswitch stops and starts the movement, but there are safety switches at the end to prevent overriding. For safety the microswitch is on a low voltage circuit operating the main motor switch by a relay. The device is also valuable for going downstairs as it prevents the person falling forward and gives a great feeling of security.

8.4.5 *Exoskeleton for arthritics*

People with arthritis often experience considerable pain from standing or walking; for example, when doing housework. This pain is due to pressure on the damaged joint, and the device shown in Plate XIV was developed as a possible aid for such conditions. It is so adjusted that most of the person's weight is taken on the bicycle saddle, through the exoskeleton, to the metal soles under the feet. Thus, the amount of weight on the joint is greatly reduced, but normal exercise, and use of the muscles, is available. This is equivalent to an exercise bicycle, but can be used during normal life. The joints are arranged to run approximately through the person's joints, the hip joint being made by means of a 'T'-shaped bearing, with two rollers operating in a curved rail, so that a virtual bearing, working through the hip joint, is obtained. The device would, in practice, be made much less conspicuous by using moulded fibre-glass. On the design shown in the figure there is no sideways movement of the legs available (standing astride), but this could be incorporated with another bearing.

8.4.6 *Powered artificial legs and arms*

A powered artificial arm with a complete set of movements of the ordinary human arm, except that there is only one fingertip movement, has been developed in the USA (the Rancho Arm*). This can be controlled by the movements of the patient's eye using voltage pick-ups.

* M. L. Moe, 'Rancho Arm', *CISM-IFToMM Symposium on the Theory and Practice of Robots and Manipulators*, 1973 (Springer Verlag, Berlin).

An active exoskeleton has been developed for paraplegics at the Institute M. Pupin, Belgrade, Yugoslavia, described by M. Vukobratovic*. This is controlled by a computer, and enables the muscles of the paralysed limbs to be exercised when the patient is strapped in to it. A walking system which bends its knee to pick its feet up is shown in Plate XV.

8.4.7 *Devices for blind people*

A great deal of work has been done to enable blind people to walk more freely under town pavement conditions. This includes walking sticks which give an audible signal with the frequency related to the distance of the stick from the object at which it is pointed, and special spectacles with a similar function. The human eye receives an extraordinary amount of information, and it is very unlikely that it will ever be possible to produce an artificial device giving more than a minute fraction of this information, but studies have been made of artificial devices implanted into the brain to give some kind of two-dimensional picture, and of devices using an area of the skin to receive two-dimensional information. A blind reading device has been developed, whereby a page can be scanned and the shape of the letters under the scanner converted into an enlarged shape which can be felt by the finger. Such devices are naturally very expensive but this would seem to be an area in which the thoughtfulness of the engineer could make a real contribution to people with a incurable deficiency.

On a much lower technology level, a blind person living alone has a problem when pouring tea into a cup to know when it is full, without wetting their finger. We developed a small device which could be hooked over the side of the cup to sound a bell when the wires are immersed. A rather simpler solution is to float a clean ping-pong ball in the hot tea so it can be felt by the hand.

8.4.8 *Devices for handicapped people living alone*

David Somerset of 'Motor Riser Chairs Ltd' has developed chairs which enable people with leg defects to get up or sit down with a reduction of 75 per cent of the forces on the joints. He also describes a power-driven seat lifter which can be combined with an ordinary lavatory seat to enable disabled people to use it alone.†

* M. Vukobratovic, 'Dynamic control of anthropomorphic active mechanisms', *CISM-IFToMM Symposium on the Theory and Practice of Robots and Manipulators*, 1973 (Springer Verlag, Berlin).

† V. Wright, B. B. Seedhom, M. Ellis and A. Amis, 'A comparative study of the knee forces during the activity of rising from a chair without the aid of arms and by the aid of a motor riser chair', Leeds University School of Medicine, June 1977.

Other problems experienced by handicapped people living alone include the inability to support the weight of a teapot or kettle when making or pouring out tea. Plate XVI shows a device to overcome these problems.

We have worked on devices to assist people to get in and out of baths by themselves. The device shown in Plate XVII is hand operated, using the bath water as the hydraulic fluid, so that a person can gradually lift themself up until they are level with the side of the bath. Another one has been developed with low voltage battery operation.

People with weak hands sometimes have difficulty in opening tins, and a battery operated tin opener has been produced, but an improvement on the lever operated devices, to enable them to be worked very slowly by people with weak hands, would be a preferable solution.

8.4.9 *Page-turner*

Many devices have been made for turning the pages of books held up so that people who cannot use their arms can read. Here again the skill of the human fingers in separating pages and turning them over one at a time, or in finding one's place in a book, is extremely hard to imitate mechanically. However, these devices can be of real value to quadriplegics who are paralysed from the neck down and have to control the device by a mouth operated blow–suck device, or by eye movement. They need help to get the book set up, but it is a great benefit if they can turn the pages without help.

8.4.10 *Emergency call devices*

Another very important problem to which much thought has been given is that of the elderly person living alone who has a sudden heart attack or other incapacity. If there is no way of warning neighbours or relatives they may be found dead weeks later. One solution that has been proposed is an emergency call system which can be operated by touch anywhere round the walls, a few inches up, but this would not deal with complete loss of consciousness. Another system which has been devised is one whereby an alarm is sounded if the lavatory chain is not pulled for twenty-four hours. A recent Japanese system* makes use of the communal television aerial system in a block of flats to carry an alarm signal from any flat when the warning button is pressed. In this way it is not necessary to put in separate warning circuits, and, by giving each alarm switch a characteristic electrical resistance, it is possible to have an indicator which shows in which flat the emergency has occurred.

* 'TV helps to sound the alarm', *New Scientist*, 26 January 1978, p. 221.

Another device* is attached to a wrist watch. If the wearer needs help they extend a miniature aerial and press a button which actuates an alarm on a receiver up to 100 m away.

8.5 SENSORY AIDS

Devices for blind people have already been discussed in section 8.4.7, but there is also a great deal of work on devices for deaf and dumb people. Hearing aids, in the form of electronic amplifiers, are well developed and many people use them, and in some cases the sound is transmitted through the bone structure rather than through the ear. However, there are people for whom such devices cannot produce an audible signal because they are so profoundly deaf. For these people there have been a number of studies on devices. At the Applied Physics Laboratory of the University of Washington† a team is working on a belt worn round the abdomen which contains 288 tiny electrodes which deliver pulses to the skin. Speech waves are converted electronically into these pulses, after processing by a computer, and it is hoped that the deaf person will be able to learn to interpret the signals as speech.

Other work† has converted speech into a frequency analysis and produced a spectrogram where the sibilant 's' sound shows up at a frequency of around 6000 Hz. A more sophisticated device makes a computer analysis of the speech wave-form to obtain the shape of the vocal tract so that a deaf person can tell whether they are making the right sound. An experiment is going on at the present time with the 'Palantype' shorthand keyboard machine with which a whole syllable can be typed simultaneously using three fingers and the typist can keep up with a 200 word per minute speaker. This was installed in the House of Commons, with a skilled operator in the Press Gallery, for a totally deaf MP, to whom the output from the machine was communicated on a screen. Such a system is clearly only available for a VIP, as it requires a full-time operator.

Toby Churchill of Cambridge was taking an electrical engineering course when he was stricken by a rare disease which left him unable to speak and with the use of only one hand, and no legs. He has developed an instrument which he calls the 'Lightwriter'. This is a typewriter with a standard keyboard where the message appears on a single line plasma display which moves across, so that one can read his reply as one talks

* 'Keeping a watch on the old folk' *New Scientist*, 26 January 1978, p. 218.

† A. F. Newall, 'Communication aids for impaired speech and hearing', *Electronics and Power*, October 1977, p. 821.

to him. He has also made an adaptation to a car so that he can drive it quite safely with his one hand, and he has adapted a hand-operated wheelchair so that he can work both wheels with one hand together, or one at a time for steering.

8.6 HOSPITAL EQUIPMENT

8.6.1 *Diagnosis*

The applied physicist has given the doctor very powerful tools for diagnosis of growths inside the body, without surgery. X-ray tomography* was invented in 1967 by G. N. Hounsfield, and since its commercial introduction in 1972 over 1000 systems have been installed in hospitals. In this system a fine pencil of X-rays passes through the plane of the patient's body to be investigated and is observed by a detector on the other side. By traversing the whole plane with this pencil and then making another traverse, it is eventually possible to produce a series of absorption readings which can be put on to a computer to give an exact map of absorptivity of each small area in the two-dimensional cross-section. This shows up the presence of tumours, for example, very much more clearly than a conventional X-ray picture, which is a projection of a three-dimensional system on to two dimensions.

Another method of whole body scanning is the use of ultrasound†. In the case of ultrasonic waves a single transducer is used which both sends out the signal and measures the reflected wave. The waves are reflected by an interface within the body and it is necessary to use a very high gain (60 dB) to detect echoes from weak interfaces. The reflections from a signal are converted into spots across a screen at positions corresponding to the interfaces so that, by scanning from a number of places, a two-dimensional pattern of the interfaces can be built up. The method is considered to be safe and painless and interfaces between soft tissues can be detected easily. Recordings can be taken of the motion of the heart valves to show the rate and amplitude of the movement.

The latest device‡ to be studied uses nuclear magnetic resonance. The tissue to be investigated is placed in a magnetic field and radio waves stimulate transitions between the spin states of the protons. It should be possible with this method to produce moving images, but it requires a

* E. Horn, 'X-ray computed tomography', *Electronics and Power*, January 1978, p. 36.

† R. Raulton and A. Shaw, 'Diagnostic ultrasound in medicine', *Engineering in Medicine*, 1978, **7**, 47–50.

‡ Ros Herman, 'NMR makes waves in medical equipment companies', *New Scientist*, 12 January 1978, p. 67.

very large magnet to put a whole person inside the magnetic field. The system is mainly detecting the presence of hydrogen in water, but also in fat. The advantages are that it does not cause tissue damage like X-rays, or body heating, like diathermy, and that it gives information of a moving character instead of a purely static picture. A miniature electronic pulse rate indicator* has been developed which converts the mechanical pressure changes of the pulse, transmitted, into electric voltage pulses; the interval between these gives the rate.

The photograph of the whole body by infra-red enables surface areas of abnormally low or high temperature to be detected, and this also has proved a useful device for diagnosing local abnormalities.

Plate XVIII shows a device built for a doctor at The London Hospital to see whether vibrating the patient in a chair at various frequencies would help to clear blocked breathing passages.

8.6.2 *Intensive care*†

After serious operations or bad accidents it is sometimes necessary to maintain continuous observation on certain functions of the patient so that urgent action can be taken if anything deteriorates. The most important instrument is usually the electro-cardiograph (ECG) which indicates the movement of the heart. Many hospitals are fitted up with a number of such beds with the ECGs connected to a central cathode ray oscilloscope. Other instruments which can be connected permanently to a central observation point so that a single nurse can watch them all are instruments for monitoring: (i) breathing (either by means of a turbine vane, or in the case of an assisted breathing machine, by the operation of the machine itself), (ii) blood pressure, and (iii) the patient's temperature. H. Wolff, head of the Bioengineering Division of the Clinical Research Centre at Harrow, has described an ambulatory monitor which can be carried around by a patient for twenty-four hours, and store a complete record of heart action on a standard cassette‡. This greatly increases the chance of spotting any abnormality.

8.6.3 *Special beds and cots*

Plate XIX illustrates an experimental bed which was made in my department in which the supports for the lower legs, thighs, head, and back could be moved hydraulically to different angles. It is particularly

* *Engineering in Medicine*, 1978, **7**, 53.

† See Caroline Weinstein, 'Instruments for the care of the critically ill', *Electronics and Power*, January 1978, p. 31.

‡ H. Wolff, lecture given to Research and Development Society, 24 January 1978.

important when a patient is sitting up in bed to support the thighs with the reverse slope, so that they do not tend to slide down in bed all the time. The system also gives the possibility of an overall negative slope of up to 20° for draining fluid from the lungs.

8.7 ARCHITECTURE FOR CRIPPLES

Ramps for wheelchairs and special lavatories convenient for those in wheelchairs have been introduced already in many public places. There are many other situations, however, in which people concerned with buildings could usefully give much more thought to the problems of handicapped or partially handicapped people. Handles for sink and bath taps and for ovens often require considerable strength. Ovens are usually inconveniently low. There are many other devices which could be incorporated into a conventional house to make it very much more convenient for a handicapped person, such as the possibility of various types of stair climbing aids. This should be a part of the standard training of an architect.

CHAPTER 9

The engineer's responsibility

9.1 THE RELATION OF ENGINEERING TO OTHER PROFESSIONS

For the purpose of this book I am taking the title 'engineer' to include all applied scientists, architects and technologists who apply the experimental and theoretical knowledge of science to the solution of practical human problems by the development of hardware, ranging from a drug to an aeroplane. It follows from this definition that the engineer is necessarily involved in the moral problem of whether his work is producing something which improves or harms the life of human beings, and he must be concerned with this problem. The pure scientist can be amoral in the sense that he regards his work as completely disconnected from human applications, although it is becoming increasingly difficult for any branch of pure science to be separated from its practical consequences for humanity. The engineer, on the other hand, must always consider questions to which there is no clear-cut answer, such as the relation between short-term benefits and long-term disadvantages, a valuable benefit to a few people combined with a small disadvantage, such as noise or pollution, to a much larger number of people, and, finally, the most difficult question of all, should he devote his skills to constructing weapons of war? In this final chapter I shall try to give guidelines to help engineers to think about these problems and, especially, to struggle in their own conscience to find their own answers to them. It is certainly not my intention to say that anybody must think as I do, and in quoting literature I will try to give both points of view as fully as I can, although I have no doubt my own bias will show through.

The fact that there are seven professional ladders (see Chapter 3) which have to be climbed if we are to reach a stable, world-wide Creative Society, shows that the engineer, as defined above, cannot deal with the whole problem alone. The engineer is primarily concerned with the right hand ladder, and is also involved deeply in the ladders connected with food and with medicine, as discussed in Chapters 6 and 8, respectively.

However, since, as we showed in Chapter 1, engineering is the profession on which the Industrial Revolution depended most essentially, the engineer has a special responsibility for the second Industrial

Revolution, the movement to a stable equilibrium civilization in place of the perpetual growth of the first Industrial Revolution. The engineer is in fact the only person who can envisage the machines and processes which must be developed in the second Industrial Revolution if our civilization is to survive. He is also the only person who can demonstrate the nonsense of the argument used by the politicians to justify the extravagant uses of the earth's resources, namely 'when it runs out the engineers will find a substitute which is just as good'. The economist is inclined to say 'the supply is inexhaustible if we increase the price'. This is, of course, theoretically true, but since the price includes an enormous capital installation, requiring other resources and very large expenditure of energy before the substitute can be produced, it is not in practice true, because humanity could not possibly afford these capital resources*. This is why the engineer must look upon *the fossil fuels as a limited capital resource for mankind*, so that, by the time they are all used we have installed a permanent viable system.

It follows that it is the personal responsibility of every engineer, or every person considering becoming an engineer, to look at the long-term, world-wide problems discussed in the first three chapters, and as far as possible to mould their career, and their discussions with all other professions, around their own personal conclusions produced by their developed conscience, as to the way in which our civilization can survive through the next century. The engineer has, in fact, a vitally important role in the key process needed if civilization is to survive a change of public ethos. The engineer, more than anyone else, can see, and must demonstrate to other people, the engineering possibilities and limitations of man's relation with the environment. Many writers have arrived at the same conclusion; for example, an editorial in *Science* by Philip H. Abelson† says:

> One important function is that of watchdog. In exploiting scientific discoveries humanity will squander resources and unwittingly conduct profoundly important experiments on itself and on the environment . . . thus, academic scientists and the scientific societies have responsibilities that they cannot escape.

He refers to the need to create judicious public opinion by providing forums, setting up fact-finding commissions, and producing objective reports with a high level of scholarship.

At the Fourth Parliamentary and Scientific Conference of the Council

* For example, we could make an oil substitute using large amounts of nuclear electricity, hydrogen from water, and carbon from limestone.

† Philip H. Abelson, 'The social responsibilities of scientists', *Science*, 1970, **167**, 241.

of Europe, November 1975, a general declaration* was produced pointing to the need for a concerted international response to the problems of global dimensions such as food, energy, raw materials, and the needs of the less developed countries. They recommended that:

(1) European governments should take the initiative in developing global policies for science and technology, and promote the integration of technological, economic, social, political, and human considerations,

(2) governments should guard against concentrating on the technical aspects of major policy decisions, to the detriment of properly presenting the values, assumptions, and estimations of risk underlying such decisions.

9.2 STUDIES OF ENGINEERING RESPONSIBILITIES

The original Hippocratic oath of the doctors laid down a code for a high level of professional honour, responsibility to the patient, and respect for his private life†.

Canadian engineers have, for many decades, had a qualifying ritual, originally written by Kipling, which refers to the engineer's responsibility in relation to his client and fellow workers. In the case of an employed engineer this would refer rather to his employer. However, the problems discussed in the earlier chapters of this book show that the work of the engineer affects all aspects of human life, particularly:

(1) its by-products and side effects, such as noise and pollution,

(2) its effects on the destruction of the environment and wild life,

(3) its effects on future generations—particularly in using up scarce resources so that they are not available in the next century.

Thus, we can say, without any doubt, that we need a Hippocratic oath for all engineers and applied scientists which will refer to their responsibility for all the consequences of their work. A number of conferences have been held on this subject, and various oaths have been put forward. I will give some of the more outstanding examples.

* A. Bouloche, K. Nichter and K. Warner, *The Sciences and Democratic Government*, 1976 (Macmillan, London).

† Lord Haden-Guest, 'Ethical problems of the scientific worker', *Nature*, 1951, **168**, 1112.

9.2.1 *Santa Clara Declaration**

The social responsibility for Engineers

The undersigned Engineers and physical and social scientists, participants in the National Science Foundation Conference on 'Engineering and the Technological Society' at the University of Santa Clara, California, are concerned with the harmful impact of technology on environments, culture and values. We are particularly disturbed by the failure of Engineers and other professionals to exercise significant responsibility for the social consequences of technology.

Therefore, we resolve to work to develop in ourselves, our students, and our colleagues a greater sense of responsibility for the implications of their efforts.

We call on all Engineers and engineering societies to join us in this urgent endeavour.

This was signed by thirty-one professors of engineering, almost all from the USA.

9.2.2 *IEEE Committee on Social Implications of Technology*†

Eighteen fundamental canons of ethics are given, of which the first three relate to these broader issues, the others to normal professional responsibilities.

Fundamental canons of ethics

(1) The engineer has proper regard for the safety, health and welfare of the public in the performance of his professional duties, and he will regard his duty to the public welfare as paramount by notifying the proper authority of any observed conditions which endanger public safety and health.

(2) The engineer does whatever is practicable to assure the safety and reliability of products for which he is responsible and accepts responsibility for personal errors.

(3) The engineer makes a reasonable effort to inform himself as to the possible consequences, direct and indirect, immediate and remote, of projects he is working on.

9.2.3

H. R. Hrubecky‡ in his paper 'The scientific and technological revolution's demand upon education for the future world society' said:

Be it resolved that, among the nations of the world the *ethos of stewardship* with its implications for prudent judgement in the management of the resources of

* *Technology and Culture*, April 1970, **11,** 241.

† S. Unger, 'Codes of Engineering Ethics', *IEEE CSIT News Letter*, Issue No. 5, December 1973.

‡ Paper presented at UNESCO Scientific and Technological Revolution Conference, Prague, 6–10 September 1976.

nature *supplant the corrupted ethos of 'control and dominion' over nature by man.* The collective response and recognition of world leadership, that indeed, this is in the best long term interests of mankind itself.

9.2.4 *The Mount Carmel Declaration on technology and moral responsibility**

This was prepared at a conference in December 1974 and has been signed by over 300 people from all over the world.

(1) We recognize the great contribution of technology to the improvement of the human condition. Yet continued intensification and extension of technology has unprecedented potentialities for evil as well as good. Technological consequences are now so ramified and interconnected, so sweeping in unforeseen results, so grave in the magnitude of the irreversible changes they induce, as to constitute a threat to the very survival of the species.

(2) While actions at the level of community and state are urgently needed, legitimate local interests must not take precedence over the common interest of all human beings in justice, happiness and peace. Responsible control of technology by social systems and institutions is an urgent global concern, overriding all conflicts of interest and all divergencies in religion, race or political allegiance. Ultimately all must benefit from the promise of technology, or all must suffer—even perish—together.

(3) Technological applications and innovations result from human actions. As such, they demand political, social, economic, ecological and above all moral evaluation. No technology is morally 'neutral'.

(4) Human beings, both as individuals and as members or agents of social institutions, bear the sole responsibility for abuses of technology. Invocation of supposedly unflexible laws of technological inertia and technological transformation is an evasion of moral and political responsibility.

(5) Creeds and moral philosophies that teach respect for human dignity can, in spite of all differences, unite in actions to cope with the problems posed by new technologies. It is an urgent task to work toward new codes for guidance in an age of pervasive technology.

(6) Every technological undertaking must respect human rights and cherish human dignity. We must not gamble with human survival. We must not degrade people into things used by machines; every technological innovation must be judged by its contributions to the development of genuinely free and creative persons.

(7) The 'developed' and the 'developing' nations have different priorities but an ultimate convergence of shared interests.

* 'Ethics in an age of pervasive technology', Technion Institute of Technology.

For the developed nations: rejection of expansion at all costs and the selfish satisfaction of ever-multiplying desires—and adoption of policies of principled restraint—with unstinting assistance to the unfortunate and the under-privileged.

For the developing nations: complementary but appropriately modified policies of principled restraint, especially in population growth, and a determination to avoid repeating the excesses and follies of the more 'developed' economies.

Absolute priority should be given to the relief of human misery, the eradication of hunger and disease, the abolition of social injustice and the achievement of lasting peace.

(8) These problems and their implications need to be discussed and investigated by all educational institutions and all media of communication. They call for intense and imaginative research enlisting the cooperation of humanists and social scientists, as well as natural scientists and technologists. Better technology is needed, but will not suffice to solve the problems caused by intensive uses of technology. We need guardian disciplines to monitor and assess technological innovations, with especial attention to their moral implications.

(9) Implementation of these purposes will demand improved social institutions through the active participation of statesmen and their expert advisers, and the informed understanding and consent of those most directly affected—especially the young, who have the greatest stake in the future.

(10) This agenda calls for sustained work on three distinct but connected tasks: the development of 'guardian disciplines' for watching, modifying, improving and restraining the human consequences of technology (a special but not exclusive responsibility of the scientists and technologists); the confluence of varying moral codes in common action; and the creation of improved educational and social institutions.

In his Joseph Wunsch lecture, 'Reflections on the Mount Carmel Declaration, 1975', Alvin M. Weinberg said that he had certain reservations when he signed the Declaration of the three following issues.

(1) Can we survive, yet avoid gambling with survival? Here, he was concerned with the fact that our present civilization is based on the ever-increasing use of energy and concludes that the development of world-wide nuclear energy and other high technology is essential for survival.

This is the argument which I have endeavoured to meet fully in Chapter 4 by showing that a much more satisfactory future for humanity can be obtained by ceasing to be 'hooked' on enormous energy consumptions.

(2) Are Social Fixes safer than Technology Fixes? He queries whether social intervention in the form, usually, of legal compulsion, may not have worse side effects than a 'Technological Fix', and he advocates energy parks, that is, enormous clusters of nuclear facilities operated by a cadre of highly skilled permanent workers.

What I have tried to show in this book is that social compulsion is no more a satisfactory future for humanity than is the concentration of all the world's limited capital resources in a few central nuclear parks from which energy radiates, in electric cables or pipes carrying hydrogen, to the rest of the world. The one solution destroys human freedom and the other certainly could not solve the problem of provision for the needs of 8000 M people. The only feasible solution is Humane Technology.

(3) Who will guard the guardians? He claims that when scientists speak outside the boundaries of their knowledge of exact science the debate is inevitably won by the scientist who warns against presumptive dangers even though he cannot verify that danger really exists.

I am afraid that all my experience goes the other way. For example, lead is still being put into petrol in spite of many warnings of its effect on human health. However, I agree with his proposal that the scientific and engineering societies should act as quasi judicial bodies to hear both sides of the case when matters outside the exact sciences are the cause of disagreement among applied scientists.

9.3 THE ENGINEER AND WAR

There are two extreme positions, with a whole range of intermediate ones taken by engineers. The one extreme is that expressed by Sir Ernst Chain*. He says:

> Capable scientists are therefore the most precious asset which a nation possesses to give it superiority over its enemies and victory or defeat is in their hands.
>
> No society can permit its scientists to back out in case of an emergency when its very life is threatened, and no scientist is morally justified when called upon in such an emergency to play his part, to deny his services to the nation of which he forms part and which defends him and his family . . . the unilateral refusal of the western scientists to participate in war technology is a step tantamount to withdrawing from the turmoil of life. . . . The first responsibility of the scientist is to the nation of which he is a member.

* Sir Ernst Chain, *Social Responsibility and the Scientist*, 1972, General Studies Project, The Schools Council (Longman, London), p. 3.

The opposite extreme is that of the Quakers.

> We utterly deny all outward war and strife and fighting with outward weapons, for any end, or under any pretence whatever; this is our testimony to the whole world*.

H. Bondi† says:

> Basically it seems to me that a citizen either has to agree or to disagree with his Government's defence policy. If he disagrees then it is his duty to make this plain, to think through the alternative he proposes and to attempt to gather support for his views in the hope of changing the policy.

I do, in fact, disagree both with the export of weapons from Britain and with the expenditure of such a large fraction of our engineering research effort and manufacturing skills on 'defence'. I believe that if we were to devote even half this effort to, for example, finding out how to help the poorer people of India to grow more food, we could, in ten years, make a far greater contribution to averting the Third World War than we shall ever do with our present policy.

There is no doubt that ever since the dropping of the nuclear bombs on Japan the applied scientist has had to consider in his conscience the terrifying consequences of the application of his skills to the destructive processes of war. The fundamental issue is whether *the individual's conscience tells him that his first loyalty is to humanity as a whole or to his own nation.* We have become strongly polarized in terms of nations, although this is a comparatively new development of the last two or three centuries even in Europe, and much newer than that elsewhere. Many people in many parts of the world give their loyalty to a religion or to a political dogma rather than to a nation, but even in these cases they would tend to put their own nation first if the transnational group were under stress. The fundamentally new features of the situation, which make it impossible to apply historical ideas, are that technology has shrunk the world and has given the 'hawks' weapons, which, if used, will ensure the total wiping out of the enemy civilization. Some people argue that this latter point is a good thing since no major power would dare to declare an all-out nuclear war or even use the most potent chemical and biological weapons, because their own civilization would also inevitably be destroyed by the opponent.

* Declaration to Charles II 1660, quoted in the Preface to *Quakers Visit Russia* by Dame Kathleen Lonsdale, in Kathleen Lonsdale's obituary, *The Times*, Friday 2 April 1971. It says 'To the Quaker conscription for military training is an offence against the human spirit. What service the individual may voluntarily undertake is a matter of conscience.'

† David Morley, *The Sensitive Scientist*, 1978 (SCM Press, London), p. 77.

This argument is, unfortunately, useless, because the 'hawks' in both countries constantly demand increasing research on more and more sophisticated weapons by arguing that the other side might get so great a lead that a first strike would prevent a counter-attack being launched, or might become so powerful that they would dare to attack with conventional weapons, hoping to win the war without the use, on either side, of the nuclear weapons. Thus, the proportion of the wealth of the rich countries being spent on 'defence' rises steadily and, in a number of countries, the whole economy depends more and more on the exports of sophisticated weapons.

In my opinion, the engineer with a conscience will find himself, or herself, much closer to the view of the Quakers than to that of the 'hawks', and much more concerned with the long-term future of humanity than with local patriotism. There are two reasons for this.

One is that to devote even a few year's of one's working life to the improving of weapons of destruction must inevitably have a terribly damaging effect on the humanity of the engineer so that he begins to think of people and their homes as targets. I believe that this is bound to be something which at the end of his life he will regret deeply. Secondly, what I have tried to show in this book is that arms escalation cannot conceivably lead to a stable twenty-first century for humanity, whereas the gradual diversion of the research effort and financial investment from weapons to Humane Technology could lead to a better life then than we have at present, both in the developed and in the underdeveloped countries. The ultimate decision must be for the conscience of each individual, but that conscience must be aware of, and strongly influenced by, the long-term world-wide considerations.

9.4 ADVICE TO YOUNG ENGINEERS AND SCIENTISTS

Anyone who has not yet decided on their career but is interested in science and technology may feel impelled not to enter a career in applied science because they are impressed by the harmful by-products of engineering. Indeed, I believe that one of the reasons for the great increase in the number of people who read social science ten years ago was the feeling of the young that they wished to make some social contribution, and that this was the best way to do it. My answer in all these cases is that we need engineering, and that the only people who can produce machines without these harmful by-products are engineers with a well-developed conscience. Similarly, I have been asked by chemical graduates whether they should refrain from entering the chemical industry because of the pollution effects of that industry. Again my answer is 'You are just the people who should enter that industry and keep your conscience strong and clear, so that as you rise up within it you can play an increasing role in eliminating pollution'. Most of the big chemical companies have anti-pollution officers, and as raw materials become shorter it will be more and more economical to recycle the pollutants instead of throwing them out in air or water, or on to tips in barrels. Research on how to do these things as economically as possibly is one of the most important lines for the chemical industry.

I can, therefore, sum up my advice to young people with a conscience who are interested in science and technology by saying: *Qualify as well as you can, get inside the system, and keep your conscience active as you follow your*

TABLE 9.1 THE MORAL SPECTRUM OF ENGINEERING

↑ Increasing value to individuals	*Machines that increase the possibilities of the individual for creative self-fulfilment.* Medical engineering and sceptrology; hygiene and sanitation; education and travel machines; communication machines; equipment for arts and crafts.	Humane Technology
	Machines that assist the individual to earn a good living without drudgery or danger at work. Telechirics (mining, undersea and in factories); agricultural machinery, 'mechanical cow'; fridges; transport. (Automation and robots only to eliminate danger or total repetition.)	
	Cosmetic machines. Non-essential status symbols, fashion engineering, and built-in obsolescence. 'Throw-away' goods; unnecessarily large or fast cars; gadgets; 'keeping up with the Joneses'; space race; supersonic flight; very large power stations.	
0	Neutral value	
Harmful ↓	*Evil by- products of cheap and thoughtless engineering.* Noise and ugliness; pollution, accidents; stress; fear; fuel and mineral exhaustion; unemployment; repetitive work.	
	Intentionally harmful. Weapons to destroy people and homes; torture machines.	

career. The only limitation is that I personally would advise you *not to go into industries where you will be concerned with making weapons or materials of war and destruction*.

9.5 THE ENGINEER'S CONSCIENCE

What I have tried to show in this book is that the engineer must not wear blinkers; he must be concerned with the long-term world-wide consequences of his engineering, and he has a vitally important role to play in educating the public to appreciate the engineering possibilities and dangers of the future. He must realize that *the subjective qualities of human life, such as self fulfilment, happiness, inner freedom, and love, have much more real long-term importance to the people affected by his engineering than does the possession of goods and status symbols beyond those necessary for a full life*. One way of expressing this is given in Table 9.1, which puts all the work of the engineer on to a moral spectrum in its relation to the life of the individual. At the top come all machines which are designed to increase the possibility of the individual for a life of creative self fulfilment. Immediately below that comes the category of machines to give a fully adequate standard of living to all people, the next category is things which are non-essential but not harmful in themselves, except in so far as they have harmful side effects. These harmful side effects, together with machines deliberately developed for harming human beings, are put below the line at the bottom of the table.

I conclude with the following, which is my own Hippocratic oath for engineers and applied scientists*.

AN OATH FOR APPLIED SCIENTISTS AND ENGINEERS

I vow to strive to apply my professional skills only to projects which, after conscientious examination, I believe to contribute to the goal of co-existence of all human beings in peace, human dignity, and self-fulfilment.

I believe that this goal requires the provision of an adequate supply of the necessities of life (good food, air, water, clothing and housing, access to natural and man-made beauty), education and opportunities to enable each person to work out for himself his life objectives and to develop creativeness and skill in the use of the hands as well as the head.

I vow to struggle through my work to minimize danger, noise, strain or invasion of privacy of the individual, pollution on earth, air and water, destruction of natural beauty, mineral resources and wild life.

* M. W. Thring, 'Our Responsibility to Mankind', p. 9 in E. J. Semmler (Ed.), *The Engineer and Society*, 1973 (Institution of Mechanical Engineers, London).

SUBJECT INDEX

AUTHOR INDEX

UNIVERSITY
OF BRISTOL
PHYSICS